Handel und Internationales Marketing
Retailing and International Marketing

Series Editors

Bernhard Swoboda, Universität Trier, Trier, Germany

Thomas Foscht, Karl-Franzens-Universität Graz, Graz, Austria

Hanna Schramm-Klein, Lehrstuhl für Marketing, Universität Siegen, Siegen, Germany

Die Schriftenreihe fördert die Themengebiete Handel und Internationales Marketing. Diese charakterisieren – jedes für sich, aber auch in inhaltlicher Kombination – die Forschungsschwerpunkte der Herausgeber. Beide Themengebiete werden grundsätzlich breit aufgefasst; die Reihe bietet sowohl Dissertationen und Habilitationen als auch Tagungs- und Sammelbänden mit unterschiedlicher inhaltlicher und methodischer Ausrichtung ein Forum. Die inhaltliche Breite ist sowohl im Sinne eines konsumentenorientierten Marketings wie auch einer marktorientierten Unternehmensführung zu verstehen. Neben den Arbeiten, die von den Herausgebern für die Schriftenreihe vorgeschlagen werden, steht die Reihe auch externen wissenschaftlichen Arbeiten offen. Diese können bei den Herausgebern eingereicht und nach einer positiven Begutachtung publiziert werden.

The book series focuses on the fields of Retailing and International Marketing. These two areas represent the research fields of the editors—each of them as a single research area, but also in combination. Both of these research areas are widely understood. Consequently, the series provides a platform for the publication of doctoral theses and habilitations, conference proceedings and edited books, as well as related methodological issues that encompass the focus of the series. The series is broad in the sense that it covers academic works in the area of consumer-oriented marketing as well as the area of marketoriented management. In addition to academic works recommended by the editors, the book series also welcomes other academic contributions. These may be submitted to the editors and will be published in the book series after a positive assessment.

Jan-Lukas Selter

Acceptance and Resistance of Innovative Technologies

Investigating the Acceptance and Resistance of Innovative Technologies with a Focus on Ethics and Sustainability

Springer Gabler

Jan-Lukas Selter
University of Siegen
Siegen, Germany

Jan-Lukas Selter, Dissertation, Universität Siegen, 2025.

ISSN 2626-3327 ISSN 2626-3335 (electronic)
Handel und Internationales Marketing Retailing and International Marketing
ISBN 978-3-658-48593-1 ISBN 978-3-658-48594-8 (eBook)
https://doi.org/10.1007/978-3-658-48594-8

This Springer Gabler imprint is published by the registered company Springer Fachmedien Wiesbaden GmbH, part of Springer Nature.
The registered company address is: Abraham-Lincoln-Str. 46, 65189 Wiesbaden, Germany

Acknowledgments

It is the day after defending my dissertation, and as I sit at my desk, I still cannot wrap my head around everything that has happened. For the past four and a half years, I have dedicated myself to this project—years full of research, writing, countless late nights, conference deadlines, and journal revisions. But beyond the work itself, it is the people who have made this time truly special.

First, I would like to thank my supervisor, Prof. Dr. Hanna Schramm-Klein, for her constant support, valuable feedback and the freedom she gave me to shape this dissertation on my own terms. Her thoughtful and respectful guidance created an environment in which I could truly grow—both personally and academically.

I would also like to sincerely thank my second examiner, Prof. Dr. Volker Stein, for his kind support and for ensuring a smooth and efficient examination process. In addition, I would like to thank Prof. Dr. Arnd Wiedemann for chairing the examination committee and creating a warm and welcoming atmosphere during the defense.

I owe a lot to my amazing colleagues, who turned work into friendship and made this whole chapter unforgettable. Huge thanks to Carmen Richter—always cheerful, the heart of the team, and never short of a great travel tip. Special thanks also go to Dr. Katja Wagner, Dr. Anne Fota, Julian Schmitz, and Dr. Tobias Röding, who have supported me from the beginning and whose valuable input has significantly shaped my research. I am also incredibly grateful to all the other colleagues who have been part of this journey and have contributed in so many meaningful ways—whether through thoughtful conversations, shared laughs, or simply being there when it counted. Eric Schell, Theresia Mennekes, Dr. Gerhard Wagner, Dr. Robér Rollin, Dr. Manuel Froitzheim, Dr. Jonas Brühl, Timo Jenne, Alessia Jahns, and Maria Bergmann—thank you all for your support, inspiration, and for making this time so rich and memorable.

A very special thank you goes to my wonderful parents, Ulla and Peter Selter, who have supported every decision I have made and laid the foundation for everything I have achieved with their unconditional love and honesty.

Finally, my fiancée, Kim Aileen Wenzel—thank you for taking this journey with me, for your patience and for your love. I am endlessly grateful to be exploring life and the world with you every single day.

Jan-Lukas Selter

Contents

Abbreviations

AI	artificial intelligence
ANOVA	Analysis of variance
app	application
AVE	average variance extracted
B2C	business-to-consumer
CEP	Courier-, Express- and Parcel
CO_2	carbon dioxide
CR	composite reliance
ELCV	electric light commercial vehicle
F	F-statistic
H	hypothesis
IoT	Internet of Things
IP	interview partner
IRT	Innovation Resistance Theory
M	mean
MS	mean squares
N	number of sample size
n.s.	not significant
p	p-value
p.	pagc
R	Pearson's r (correlation coefficient)
R^2	R-squared (coefficient of determination)
RQ	research question
SD	standard deviation
sig.	significance level
SPSS	Statistical Package for the Social Sciences
t	t-statistic

TBL	triple bottom line
UTAUT	Unified Theory of Acceptance and Use of Technology
α	Cronbach's alpha

List of Figures

List of Tables

Introduction 1

1.1 Motivation of the Dissertation

The dissertation examines the adoption of different innovative technologies and explores the behavioral intention of both consumers and companies of using these technologies. Both determinants of acceptance and resistance of technology are considered by referring to the Unified Theory of Acceptance and Use of Technology (Venkatesh et al., 2003) and the Innovation Resistance Theory (Ram and Sheth, 1989). The six essays in this dissertation relate to six innovative technologies and approaches. This entails an examination of technologies that either involve ethical challenges or are designed with the objective of achieving (ecological) sustainability. Firstly, the ethical dimensions of artificial intelligence (AI) are examined, and a comprehensive overview of this topic is presented. Furthermore, the examination extends to automated purchasing processes in a consumer context and the role of trust and control for using these processes. The relevance of trust and trust-related factors is also considered in the context of the use of cryptocurrencies, representing another innovative technology. In the context of online retailing, two innovative approaches are examined that are intended to reduce returns, contribute to ecological sustainability, and improve consumers' online retailing experience. This includes different approaches to digital nudging as well as online size decision aids. Challenges such as data disclosure and complexity are considered in this context. Finally, another goal of this dissertation is to investigate sustainability in logistics after the online purchase, more precisely on the last mile, from the perspective of the various stakeholders. Accordingly, electric vehicles are investigated as the sixth innovative technology.

J. Selter, *Acceptance and Resistance of Innovative Technologies*, Handel und Internationales Marketing Retailing and International Marketing,
https://doi.org/10.1007/978-3-658-48594-8_1

Technological advancement has accelerated to unprecedented rates, exerting disruptive power (Chui et al., 2024). The integration of innovative technologies and systems has become integral to daily life and no longer represents mere theoretical concepts (KPMG LLP, 2023; Rejeb et al., 2023). PricewaterhouseCoopers has identified eight essential technologies that are perceived to be most transformative and fundamental for everyday operations and long-term strategy. These include AI, Internet of Things (IoT), and blockchain (PricewaterhouseCoopers, 2024b). With the technological landscape continuing to evolve, the integration of such cutting-edge technologies has become a central driver of global trends, influencing the behavior and opportunities of consumers (Hoffman et al., 2022) and the operations of businesses (Chui et al., 2024). All sectors are susceptible to technology-based transformation and change, with retailing being significantly influenced for example by advanced connectivity and applied AI (Chui et al., 2024). AI occupies a pivotal position in the context of innovative technologies. It is not only an innovative technology by itself but also serves as a catalyst for the development and deployment of other applications (Guha et al., 2021; McKinsey & Company, 2024). In particular, generative AI—AI that can create new content—is employed by consumers and perceived to be beneficial in various application areas (KPMG LLP, 2023). It is estimated that the use of generative AI can generate significant economic value up to 4.4 trillion US dollars annually (McKinsey & Company, 2024). AI facilitates the implementation of further innovative applications, including automated and personalized processes (Hoffman et al., 2022). In addition to AI, other technologies are also attracting attention. For example, companies are increasingly planning to invest in the IoT and blockchain (PricewaterhouseCoopers, 2024a) and consumers are increasingly recognizing the utility of these technologies. For instance, 71% and 76% of consumers, respectively, view AI and IoT as at least little useful, while at least 53% view cryptocurrency, which is based on blockchain, as somewhat useful (KPMG LLP, 2023).

Another key trend is the electrification of the mobility sector. There is a notable interest among businesses in this technology, with huge investments being made to achieve reductions in global carbon emissions (McKinsey & Company, 2024). The popularity of electrified vehicles has resulted in a high number of sales. However, there are still challenges, such as high pricing and range anxiety, which are impeding the proliferation of electric vehicles (McKinsey & Company, 2024). The intensive use of emerging technologies by companies results in various positive effects in the consumer context, including improved service quality, optimized decision-making, and an increase in cyber security, risk management, and transparency (PricewaterhouseCoopers, 2024a).

However, challenges and concerns such as data protection and security concerns also emerge (KPMG LLP, 2023). One key concern in this context is the ethical dimension of innovative technologies (Attard-Frost et al., 2023). Ethics is a central concept in philosophy, exploring the principles and values that determine what is considered morally right or wrong (Stahl, 2012). These principles can be understood through distinct branches of ethics, including normative ethics, descriptive ethics, and metaethics (Stahl, 2012). Normative ethics is divided into three subcategories: consequentialism, which evaluates the morality of actions based on their outcomes (Bentham, 1948 (orig. 1789)), deontology, which assesses actions according to adherence to duties or rules (Kant, 1785), and virtue ethics, which focuses on developing good character traits that promote human flourishing (MacIntyre, 1985). These approaches can be traced back to influential philosophers such as Jeremy Bentham, Immanuel Kant, and Aristotle. Furthermore, applied ethics has emerged as a field of study to address specific real-world contexts (Fieser, 2000). Examples of such contexts include medical ethics (Beauchamp, 2003), consumer ethics (Hassan et al., 2022), and digital ethics (Ashok et al., 2022). In the context of innovative technologies, ethical issues assume a central role, as technologies like AI lead to disruptive changes and pose new challenges (Du and Xie, 2021). These ethical issues include, but are not limited to, concerns related to privacy, responsibility, transparency, and aspects of trust (Du and Xie, 2021; IBM, 2024). They emerge from various aspects like the potential threat to or restriction of human rights (Livingston and Risse, 2019). Additionally, concerns may arise regarding the responsibility for autonomous systems and technologies or the relinquishment of control (Macnish and Fernandez Inguanzo, 2019). The delegation of responsibility to technology by humans limits their own ethical responsibility (Heyder et al., 2023). It is imperative that users relinquish control to technology for an appropriate usage, however, this may result in the formation of psychological barriers (Jörling et al., 2019). In the context of consumer behavior and innovative technologies, the investigation of ethical considerations plays a pivotal role, and further research is warranted (Hassan et al., 2022). Consequently, the dissertation places a particular emphasis on the examination of innovative technologies that raise ethical challenges. Technologies that present ethical challenges and are investigated in the dissertation include AI, due to issues such as responsibility and transparency (Dwivedi et al., 2021), and automated processes, due to the pass-over of control on technologies (Jörling et al., 2019). Furthermore, the use of cryptocurrencies presents ethical challenges due to issues including concerns of trust and transparency (Mendoza-Tello et al., 2019). Additionally, digital nudges present an innovative approach being ethically problematic due to the manipulation of consumers (Weinmann et al., 2016). These are four of the six innovative technologies examined in the dissertation, which will be presented in more detail in the Sects. 1.2 and 1.3.

Another key challenge in the context of technological innovation is the growing need to integrate environmental sustainability. Environmental sustainability has become a global challenge for businesses and governments (Vadakkepatt et al., 2021; Zhang, 2023), especially as climate-related disasters continue to add costs that are expected to exceed $1.5 trillion by 2050 (Capgemini Research Institute, 2023). The relevance of environmental sustainability has become more apparent to companies in recent years (Lukin et al., 2022; Vass et al., 2021), with 63% of executives identifying sustainability as a key business case in 2023 (2022: only 21%), and many companies recognizing that the benefits of focusing on sustainability are more relevant than the costs incurred (Capgemini Research Institute, 2023). Although companies are adapting their processes accordingly, this is not yet reflected in a significant increase in investments (Capgemini Research Institute, 2023). However, regulations are forcing companies to rethink and adapt their processes. For example, the EU has implemented regulations to promote environmental sustainability, which include not only environmental but also social and economic benefits (Brennan et al., 2023). Considering sustainability and sustainable development, the 17 Sustainable Development Goals of the United Nations should be mentioned. By addressing the world's major challenges in a comprehensive and integrated manner, the goals are highly relevant, taking into account economic, social, and environmental aspects (United Nations, 2023). In the context of retailing, and with regard to the dissertation, the Sustainable Development Goal "responsible consumption and production" plays an important role and is defined by the Oslo Symposium in 1995 as "the use of services and related products, which respond to basic needs and bring a better quality of life while minimizing the use of natural resources and toxic materials as well as the emissions of waste and pollutants over the life cycle of the service or product so as not to jeopardize the needs of further generations" (IISD, 1995).

Consumers are also placing greater emphasis on sustainable consumption (Reichheld et al., 2023). Studies by Deloitte and Ernst & Young have shown that consumers in 2023 attach much greater importance to sustainable consumption and are increasingly trying to adopt sustainable consumption behaviors (Deloitte, 2024; Swallow, 2023). Especially younger consumers are interested in buying products from sustainable companies (Reichheld et al., 2023). This shows that sustainable behavior is actively practiced by many consumers, whereas sustainability was only talked about but not implemented in the past (KPMG LLP, 2023). Many technologies already exist to reduce global emissions (McKinsey & Company, 2024), and companies are beginning to use technologies to transform their products and services to support the ecological environment (Chui et al., 2024). This dissertation places a particular emphasis on these innovative technologies that contribute to improved (environmental) sustainability. One area representing an economic and environmental challenge is retailing, with high return rates in e-commerce (Rees, 2023). Innovative approaches to reduce these return rates include digital nudges (Weinmann et al., 2016) and size guides, which are already used by most online retailers (McKinsey & Company, 2021). While digital nudges might be considered ethically questionable due to their subconscious influence, decision aids such as size guides are intended to consciously assist consumers (Ghasemaghaei et al., 2019). In addition, the delivery of e-commerce orders has a significant impact on the environment (Wolff and Madlener, 2019). Sustainability concerns in logistics in general have led to technological developments in mobility and the implementation of electric vehicles (McKinsey & Company, 2024). These three innovative technologies of digital nudges, decision support tools, and electric vehicles will be explored in more detail in the dissertation along with the aforementioned technologies of AI, automated purchase processes and cryptocurrencies and will also be introduced in more detail in the Sects. 1.2 and 1.3.

1.2 Theoretical Foundation and Central Domains of Acceptance and Resistance of Technology

Research regaring the acceptance of technologies is characterized by various approaches being developed and continuously refined over the years. One of the earlier approaches that emerged in the field of social psychology is the Theory of Reasoned Action proposed by Ajzen and Fishbein (1975). The Theory of Reasoned Action represents a fundamental model for predicting human behavior, which is primarily determined by individuals' intention (Venkatesh et al., 2003). This is further influenced by the evaluation of the behavior as positive or negative and by the social pressure to perform or refrain from the behavior (Ajzen and Fishbein, 1975). The Theory of Reasoned Action is a general approach that was not developed with the specific intention of being used in the context of technology, however, it has been employed in this context by research (Momani and Jamous, 2017). Building on the Theory of Reasoned Action, Ajzen (1985) introduced the Theory of Planned Behavior, which extends the Theory of Reasoned Action with the addition of perceived behavioral control as a third influencing factor. The Theory of Planned Behavior also represents a general approach that was developed for consideration of individual behavior in a non-specific context

(Ajzen, 1991). In the context of technology acceptance, the Technology Acceptance Model is one of the most established approaches and was proposed by Davis (1989). The Technology Acceptance Model is partially based on the Theory of Reasoned Action and postulates that perceived usefulness and perceived ease of use influence attitudes toward the use of technologies and, subsequently, actual use behavior (Davis, 1989). The Technology Acceptance Model elucidates the processes through which users form acceptance or rejection of (novel) technologies, a concept that was further developed in the Technology Acceptance Model 2 by Venkatesh and Davis (2000). The Technology Acceptance Model 2 introduces additional influencing factors, thereby attempting to provide a more detailed explanation of the factors that contribute to the acceptance and use of technologies (Venkatesh and Davis, 2000).

A model that considered and included these mentioned acceptance models as well as further acceptance models in its conceptualization to create a unified approach is the Unified Theory of Acceptance and Use of Technology (UTAUT) by Venkatesh et al. (2003). The authors identified the three determinants of performance expectancy, effort expectancy and social influence directly influencing behavioral intention. They further identified facilitating conditions and behavioral intention directly influencing use behavior (Venkatesh et al., 2003). Furthermore, the model is augmented by four moderators that exert influence on the various relationships (Venkatesh, 2022). The variables of the UTAUT were developed with reference to the constructs of other established acceptance theories, making the model one of the most integrated acceptance theories (Momani, 2020). According to Venkatesh et al. (2003), performance expectancy refers to the extend to which an individual believes that using the system or technology will enhance their performance. The determinant focuses on task accomplishment. Effort expectancy refers to the level of ease and complexity associated with using the system. Social influence refers to the extend to which an individual perceives that significant others think they should use the new system, while facilitating conditions relate to the belief that sufficient organizational and technical infrastructure exist to support the technology's use (Venkatesh et al., 2003). Behavioral intention is understood as an individual's intention to utilize a technology and represents a fundamental concept within user acceptance models (Venkatesh et al., 2003). Usage intention is therefore a predictive indicator of actual use, where a stronger intention typically leads to a higher probability of actual use (Venkatesh et al., 2003). This dissertation does not investigate the resulting direct influence on the actual use of the technology. In their model Venkatesh et al. (2003) concentrated on a few selected determinants of acceptance, drawn from the various constructs of the eight theories on which the UTAUT is based. This dissertation will also adopt this approach, examining various determinants of acceptance that influence individuals' intention to use technologies. Hereby, the focus is on the investigation of other determinants than presented in the UTAUT.

In the context of marketing literature, the subject of consumer response to innovations is divided into the acceptance and resistance of these innovations (Laukkanen, 2016). Resistance to innovations is a common cause of their failure. It is defined as "the resistance offered by consumers to an innovation, either because it poses potential changes from a satisfactory status quo or because it conflicts with their belief structure" (Ram and Sheth, 1989, p. 6). This resistance occurs despite the various advantages of using the technologies (Kleijnen et al., 2009) and is addressed in detail by Innovation Resistance Theory (IRT) of Ram and Sheth (1989). IRT provides a comprehensive framework for understanding consumers' resistance to new technologies, with a particular focus on psychological factors (Frank et al., 2023). In the context of IRT, resistance can be classified into two categories: active and passive resistance. Passive resistance refers to resistance that emerges prior to the evaluation phase of an innovation and is characterised by negative preconceptions towards innovations in general (Heidenreich and Handrich, 2015b). As posited by Heidenreich and Handrich (2015b), passive resistance is most prevalent among individuals who exhibit a trend towards resisting change or who are content with the status quo. This resistance therefore also arises from a general rejection of change or from satisfaction with the existing conditions (Huang et al., 2021). Active resistance to innovations is related to a negative assessment of these innovations subsequent to evaluation (Heidenreich and Spieth, 2013). This form of resistance manifests itself in the form of negative attitudes and intentions towards innovations (Huang et al., 2021). Passive resistance occurs unconsciously and prior to any direct confrontation with the innovation, whereas active resistance represents a conscious reaction to the evaluation of the innovation (Talke and Heidenreich, 2014). Both forms of resistance can manifest themselves in various behaviors, such as a direct rejection of the product, delayed decision-making, or strong negative feelings towards the innovation (Kleijnen et al., 2009). Additionally, the literature distinguishes between functional and psychological barriers to resistance, offering valuable insights into the reasons why individuals resist new technologies (Ram and Sheth, 1989). Functional barriers refer to the perceived practical shortcomings of an innovation that make it unattractive to potential users (Talke and Heidenreich, 2014) and can be divided into usage, value and risk barriers (Talwar et al., 2020). Usage barriers pertain to the difficulties that may emerge when introducing an innovation in comparison to existing solutions (Ram and Sheth, 1989). In the case of digital innovations, in particular, the complexity of the application can represent a significant barrier to adoption (Kaur et al., 2020). Value barriers emerge when an innovation fails to align with the prevailing value system, and the costs and effort involved are perceived to outweigh the benefits (Kaur et al., 2020). Risk barriers arise when uncertainties and the associated risk impede the adoption of the innovation (Kaur et al., 2020). Psychological barriers are the result of existing user beliefs and the risks associated with them (Kleijnen et al., 2009). They can be divided into two categories: tradition barriers and image barriers (Sadiq et al., 2021). Tradition barriers emerge when an innovation necessitates changes to established routines or user behavior, leading to resistance (Kaur et al., 2020). Image barriers relate to negative sentiments towards innovations stemming from their perceived complexity (Lian and Yen, 2013). The dissertation examines the fundamental assumptions of IRT and investigates the multiple factors that contribute to resistance to innovations, ultimately deterring consumers from adopting them.

A comprehensive understanding of innovation adoption necessitates the consideration of both the acceptance and resistance of innovations (Heidenreich and Handrich, 2015b). A review of the literature on innovative technologies indicates that both IRT and acceptance theories, such as the UTAUT model, frequently offer common ground (Huang et al., 2021). For instance, a review of 54 articles by Talwar et al. (2020) demonstrated that IRT is the most prevalent theory employed for the investigation of resistance to digital innovations, with many of these studies also referring to UTAUT. This interplay of determinants for acceptance and resistance shapes consumer behavior and is examined in the dissertation and the context of six essays. As the acceptance and resistance of technologies differ with the type of innovation (Talwar et al., 2020), a range of technologies are considered in the essays of this dissertation. These theoretical considerations are presented in the conceptual framework of this dissertation, which is illustrated in Fig. 1.1. The framework forms the basis of this dissertation and is examined through individual research questions and foci in the six essays.

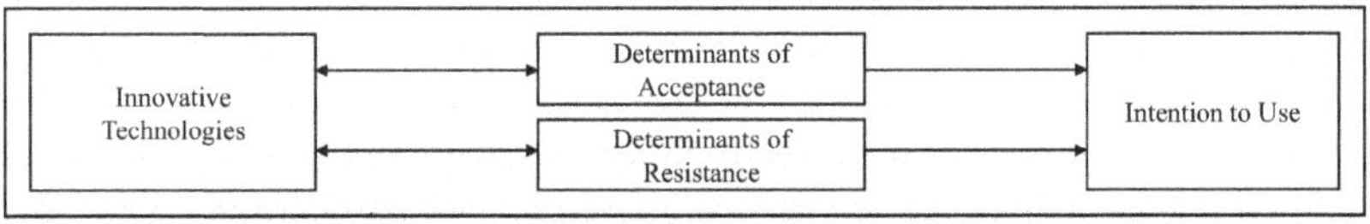

Fig. 1.1 Conceptual Framework of the Dissertation

The dependent variable that serves as the primary focus of the individual essays is the intention to utilize innovative technologies. Usage intention has been the subject of extensive examination in previous research studies, and is influenced by different determinants (Venkatesh et al., 2012). These determinants can be classified into drivers of acceptance and drivers of resistance, as referenced in the UTAUT (Venkatesh et al., 2003) and IRT (Ram and Sheth, 1989). The conceptual framework of the dissertation is primarily based on the fundamental assumptions of these two theories. The UTAUT indicates that the acceptance of technologies is influenced by a multitude of factors, which collectively shape the intention to use these technologies (Venkatesh et al., 2003). Based on this, the conceptual framework of the dissertation examines determinants of acceptance, which are broken down into various influencing factors in the essays. The IRT refers to the underlying reasons and mechanisms through which individuals exhibit resistance and rejection towards novel technologies, while also examining the various barriers that influence the decision-making process (Ram and Sheth, 1989). Based on this, the conceptual framework considers determinants of resistance, which are further subdivided into various barriers in the essays. The comprehensive analysis of both acceptance and resistance offers a holistic perspective on the adoption of new technologies. The conceptual framework is anchored in the domain of innovative technologies. While the type of innovation plays a decisive role for consumer attitudes of using these technologies (Talwar et al., 2020), the different essays of this dissertation delve into the nuances of distinct innovative technologies, striving for a comprehensive examination. The technologies selected for examination are AI, automated shopping systems, cryptocurrency, digital nudging, online decision aids, and electromobility. As illustrated in Fig. 1.1 and the conceptual framework of this dissertation, the innovative technologies are interrelated with the determinants of acceptance and resistance.

AI is transforming industries, for example by enhancing automation, personalization, and decision-making processes (Dwivedi et al., 2021). As AI becomes more integrated into daily life, it is imperative to gain a deeper understanding of the factors influencing its usage in order to understand its future impact on society. The importance of examining the factors influencing the acceptance (Kelly et al., 2023) and resistance (Hong and Cho, 2023) of AI in the context of UTAUT and IRT has been highlighted in the literature and is expanded by considering these aspects in the context of this dissertation. Another technology that is transforming the retail landscape are automated processes, such as automated buying or personalized shopping experiences (Bellis and Johar, 2020). An investigation of the factors influencing acceptance and resistance in this context can provide insights into the future of convenience-driven consumer behaviors. The successful reference to UTAUT in the consideration of autonomous technologies in the shopping context has been shown (Sharma et al., 2022) and offers further possibilities for investigation, which is carried out in the dissertation. The third innovative technology considered in the dissertation are cryptocurrencies. Cryptocurrencies represent a novel frontier in financial innovation, offering decentralized alternatives to traditional banking systems. Given the rapid rise of cryptocurrencies and the associated regulatory uncertainties, it is imperative to gain a deeper understanding of consumer behavior to use and establish this technology (Corbet et al., 2019). Research has highlighted that resistance to cryptocurrencies in the context of IRT offers valuable insights into the behavioral intention of using the technology (Sangari and Mashatan, 2024). Consequently, further examination of the factors influencing acceptance and resistance is important in this context and conducted in the dissertations. As digital platforms increasingly employ subtle techniques like digital nudging to influence decision-making, an understanding of how consumers respond to digital nudging can offer insights into ethical considerations and the effectiveness of behavioral design in online environments (Weinmann et al., 2016). Furthermore, in the context of acceptance and resistance, digital nudging represents a pivotal approach to overcoming resistance to the adoption of innovative technologies (Stryja and Satzger, 2019). Consequently, a more comprehensive examination of the underlying determinants is required in this context. Concerning the growth of e-commerce, decision aids such as recommendation algorithms have assumed a significant role in guiding purchasing decisions (Xiao and Benbasat, 2007). Prior literature in this area has only made occasionally references to UTAUT (Komiak and Benbasat, 2006) and has focused on organizational levels of innovation resistance (Mahmud et al., 2023). An examination of the factors influencing acceptance and resistance of consumers can help to understand how these technologies impact consumers' online shopping experience. The sixth innovative technology of the framework is electromobility. The transition to electric vehicles and sustainable transportation is crucial for addressing climate change (Rezvani et al., 2015). Kumar Jain et al. (2022) employed the UTAUT model to investigate the influence of determinants of acceptance on the adoption of electric vehicles. Similarly, Xue et al. (2024) employed IRT to examine the impediments to the adoption of electric vehicles. A deeper understanding of consumer acceptance and resistance towards electric vehicles is essential for forecasting adoption trends and policy support for sustainable mobility. This topic is further investigated in the dissertation.

AI, automated shopping systems, cryptocurrency, and digital nudging entail unique ethical challenges, particularly in terms of privacy, data security, decision autonomy, and fairness (Berger et al., 2022; Milano et al., 2021; Rehman et al., 2020; Wünderlich et al., 2015). These technologies rise ethical considerations regarding the responsible use of emerging innovations, as their widespread adoption has the potential to result in considerable societal impacts. Digital nudging, online decision aids, and electromobility are implemented in businesses and daily life with an increasing focus on sustainability. Main objectives of these technologies are the reduction of resource consumption, the promotion of an environmentally conscious behavior, and the facilitation of the transition to more eco-friendly systems (Berger et al., 2022; Wolff and Madlener, 2019; Yang and Xiong, 2019). It is essential to investigate the factors that influence acceptance and resistance in the context of these technologies in order to comprehend how individuals and society respond to these innovative technologies. Table 1.1 provides an overview and definitions of the selected technologies.

Table 1.1 Overview and Definitions of the Selected Technologies

Innovative Technology	Definition in the context of the dissertation
AI	The term AI is used to describe the use of computational systems that have been designed to replicate human capabilities. These capabilities include the ability to perform physical or mechanical tasks, as well as cognitive processes such as thinking and feeling (Huang and Rust, 2021).
Automated Shopping Process	The term automated shopping process is used to describe technologies that afford consumers the ability to delegate significant parts of the shopping process, including the capacity to make decisions and to perform tasks (Bellis and Johar, 2020).
Cryptocurrency	Cryptocurrency is a decentralized digital currency that leverages blockchain technology. It combines asymmetric cryptography with a distributed ledger system to ensure secure, transparent transactions without the need for intermediaries such as governments or institutions. Cryptocurrencies provide anonymous, near-instant global transfers (Wątorek et al., 2021).
Digital Nudges	Digital nudges are subtle design interventions within user interfaces that aim to influence users' decisions or behaviors without restricting their options. By leveraging elements such as default settings, prompts, or visual cues, digital nudging influences how individuals interact with online platforms, often encouraging more favorable or intended choices (Weinmann et al., 2016).

(continued)

Table 1.1 (continued)

Innovative Technology	Definition in the context of the dissertation
Online Decision Aids	Decision aids, also known as recommendation agents, are algorithm-driven tools designed to assist users by filtering, prioritizing, and suggesting products or services based on individual preferences, past behavior, or specific criteria. These systems enhance decision-making by simplifying the evaluation process, providing personalized recommendations, and improving the overall shopping experience (Ghasemaghaei et al., 2019; Yang and Xiong, 2019).
Electromobility	The term electromobility is used to describe the use of electric-powered vehicles that rely on electricity from batteries or fuel cells in place of conventional internal combustion engines. It represents a crucial element of sustainable transportation strategies, which are designed to reduce greenhouse gas emissions, diminish reliance on fossil fuels, and enhance energy efficiency in the mobility sector (Tsakalidis et al., 2020; Wolff and Madlener, 2019).

1.3 Topics of Interest and Research Questions

The dissertation identifies four topics of interest, that are listed in Table 1.2. The goal of this dissertation is to obtain a profounder understanding of these topics and the objective of the six essays is to enhance comprehension of the topics. Table 1.2 illustrates the correlation between the essays and the topics of interest.

Table 1.2 General Research Goals

Topics of Interest *A profounder understanding of…*	Essay 1	Essay 2	Essay 3	Essay 4	Essay 5	Essay 6
determinants of acceptance and resistance to use innovative technologies.	✓	✓	✓	✓	✓	✓
…consumers' intention to use innovative technologies.		✓	✓	✓	✓	
… the ethical handling of innovative technologies.	✓	✓	✓	✓		
… the use of innovative technologies to shape (ecological) sustainability.				✓	✓	✓

As previously mentioned, the intention to use innovative technologies is influenced by a number of factors that affect the level of acceptance and resistance (Ram and Sheth, 1989; Venkatesh et al., 2003). The six essays collectively examine a range of these determinants. Furthermore, four of the essays explicitly address consumers' intention to use innovative technologies as a key area of focus. In situations of uncertainty, consumers will seek to maximize utility and compare the benefits of technologies with the costs (Neumann and Morgenstern, 1947). This might lead to factors of acceptance or resistance. The dissertation's additional research objectives are to gain a deeper comprehension of the ethical challenges rising with innovative technologies and to utilize innovative technologies to promote environmental sustainability. All essays either examine technologies in which ethical challenges assume a pivotal role or which have been developed with a focus on improving sustainability. To investigate these topics of interest within the scope of the dissertation, six research questions are derived in the following and are addressed in the six essays of the dissertation. By deriving the research questions, the six innovative technologies mentioned in Sect. 1.2 are also considered in more detail.

AI offers the possibility of transforming and augmenting human tasks and activities in various industrial, intellectual, or social applications (Dwivedi et al., 2021), thereby overcoming the limitations of human intelligence and opening up new possibilities and opportunities (Daugherty and Wilson, 2018). While AI offers significant benefits, its rapid development also presents a range of substantial challenges (Mikalef et al., 2022), many of which require critical examination (Milano et al., 2021). Literature identifies different challenges associated with AI, including economic concerns pertaining to implementation costs, data security, transparency issues, and political challenges involving data rights and the risk of foreign interference (Dwivedi et al., 2021). Furthermore, ethical concerns emerge, for example with regard to the accountability for AI-driven decisions and the comparability of human and machine values (Dwivedi et al., 2021; Milano et al., 2021). The ethical implications of AI use have generated a growing interest in academic research. Literature in this field addresses a number of domains, with an increasing focus on the ethical challenges associated with the use of AI in marketing strategies, including sales processes or customer service options, as well as customer behaviors (Davenport et al., 2020). For example, discrimination against particular user groups (Libai et al., 2020), the manipulation of behavior on digital platforms (Kaplan and Haenlein, 2019), and the loss of control over AI systems (Steels and Lopez de Mantaras, 2018) are some of the ethical concerns that have been raised. Furthermore, societal barriers, such as cultural norms, influence the ethical use of AI (Dwivedi et al., 2021). The development and deployment of AI

(Umbrello, 2020), as well as the assignment of moral responsibility to AI systems (Tigard, 2021), are pivotal topics in ongoing discourses. AI systems collect a considerable quantity of user data, including both information that is knowingly provided and inadvertently left behind through the routine use of digital technologies. The disclosure of this data may be either conscious or unconscious, giving rise to concerns pertaining the transparency and control of data (Puntoni et al., 2021). The accelerated development of AI technologies has given rise to concerns that ethical issues are not being sufficiently addressed. In particular, the questions of responsibility and accountability for AI systems remain unresolved (Dwivedi et al., 2021). Morley et al. (2020) mentioned that the implementation of AI principles remains inadequate. Despite the development of ethical guidelines and principles at the policy level, their practical application remains inadequate (Puntoni et al., 2021).

The expanding corpus of research on the ethical implications of human interactions with AI has given rise to a multitude of research approaches (Heyder et al., 2023). To address these challenges, a comprehensive investigation of the ethical and moral dimensions of AI is required to establish a foundation for future research and practical insights. This dissertation aims to present a systematic review of the literature on AI ethics, identifying gaps and future research directions through the systematization of existing knowledge (Briner and Denyer, 2012). Consequently, the first research question of the dissertation is:

- How does the scientific literature address the multifaceted relationship between ethics, morality, and AI, and how can literature in this context be systematized into different categories?

Automated shopping processes offer numerous benefits, including enhanced convenience, time savings, and reduced cognitive effort in managing daily tasks (Puntoni et al., 2021; Sequeiros et al., 2022). By interacting with their environment and leveraging IoT capabilities, smart devices can predict consumer needs based on big data and execute automated processes (Jörling et al., 2019). As technological advancements continue, these systems are evolving from simple automation to autonomous purchasing decisions (Bellis and Johar, 2020). For instance, smart refrigerators are able to recognize food items and assist consumers with purchasing decisions (Bellis and Johar, 2020). As the market for IoT-enabled smart home devices continues to grow, automated shopping is rapidly gaining traction in different industries, particularly in the retail sector, with the systems streamlining consumer interactions and enhancing the overall shopping experience (Hollebeek et al., 2021; Sequeiros et al., 2022). Despite the advantages of automated shopping processes, certain challenges hinder a more widespread adoption (Bellis and Johar, 2020). These challenges are not technical but psychological, centering on how consumers perceive their role and control in these automated processes (Jörling et al., 2019). Literature indicates that consumers place a high value on their capacity to exercise control over their shopping experiences (Jörling et al., 2019). Conversely, the act of transferring this control to automated systems has been shown to evoke a sense of disempowerment (Jörling et al., 2019). This apprehension has its roots in broader societal anxieties about automation, which have been intensified by media portrayals of technology reducing human power (Puntoni et al., 2021). Control represents a pivotal psychological factor in consumer behavior. A sense of control over the environment has been highlighted to be closely linked to well-being and satisfaction in numerous studies (Leotti et al., 2010). A sense of loss of control due to a lack of transparency or limited involvement in decision-making, can rise resistance towards the adoption of automated technologies (Puntoni et al., 2021). For example, consumers may be reluctant to place trust in automated systems if they perceive exclusion from the purchasing process or if the process appears complex and challenging to comprehend (Wünderlich et al., 2013). Additionally, concerns of making erroneous decisions, which may have been influenced by factors beyond their control, contribute to the psychological obstacles against the adoption of automated shopping.

In addition to control, trust plays an important role for consumer acceptance of automated shopping technologies. Consumers are more likely to adopt these technologies if they perceive them to be trustworthy and reliable (Wünderlich et al., 2015). When automation operates in an unobtrusive manner, consumers may experience a sense of disconnection from the process, which can lead to reluctance of embracing such innovations (Kahlert et al., 2017). To address these concerns, it is imperative that automated systems provide transparency regarding decision-making processes and afford consumers the option to oversee or intervene the process when necessary (Puntoni et al., 2021). By fostering a sense of shared control and automation providing assistance rather than overriding consumer decisions, companies can mitigate perceived risks and enhance consumer trust (Bellis and Johar, 2020). The increasing body of research on automated shopping indicates that a balance must be struck between the benefits of automation, such as convenience and time savings, and the psychological need for control and transparency (Sequeiros et al., 2022). Consumers who feel empowered to oversee or intervene in automated processes are more likely to embrace these innovations, which in turn leads to higher levels of adoption (Wünderlich et al., 2015). While previous studies have examined various factors individually, few have considered how these factors interact to influence consumer behavior in the context of automated shopping processes. Addressing this gap is essential for improving the design and acceptance of these systems, leading to the second research question of the dissertation:

- What are the key factors influencing consumers' intention to use automated shopping processes and how do these factors interact to shape their willingness to relinquish control to automated technologies?

Cryptocurrencies, which utilize blockchain's decentralized and cryptographic mechanisms, are increasingly discussed and used in investment and payment applications (Sparkes, 2021). While initial research focused on the technical aspects of blockchain, the growing market capitalization of individual cryptocurrencies has shifted attention towards their economic and societal impacts (Holub and Johnson, 2018). Despite blockchain's promise of immutability and security (Feng et al., 2018), consumers often associate cryptocurrencies with uncertainty and a lack of trust (Arli et al., 2021). Trust is a crucial determinant for the adoption of most new technologies (Constantinides et al., 2017), and also mentioned as a key determinant for the adoption of cryptocurrencies (Kumar et al., 2021). Unlike traditional financial systems, cryptocurrencies face unique trust challenges due to characteristics like high volatility, decentralized structure, and limited government regulation (Holub and Johnson, 2018; Li and Wang, 2017). These issues contribute to the perceived risk and insecurity associated with digital currencies, acting as barriers to a widespread adoption (Mendoza-Tello et al., 2019; Rehman et al., 2020). For long-term success, fostering trust within the cryptocurrency ecosystem is essential (Hughes et al., 2019). Perceived risk, particularly the fear of financial loss due to price fluctuations or system vulnerabilities, is a significant deterrent (Mehta et al., 2022; Zhao et al., 2020). Similarly, concerns about the security of digital wallets and transactions as well as data security contribute to users' reluctance to engage with cryptocurrencies (Rehman et al., 2020). These factors highlight the complexity of building trust in this innovative financial technology, requring a deeper exploration of trust-related determinants.

An important strategy for addressing the trust gap in cryptocurrencies might be the transfer of trust from existing, well-established financial platforms to cryptocurrencies. According to the Trust Transfer Model of Stewart (2003, 2006), trust in an established system can influence user acceptance of newer technologies that are linked with the system. This model has been successfully applied in several domains, including payment systems (Cao et al., 2018). Additionally, research has expanded this concept by considering the transfer of other determinants, such as security and usability, from traditional systems to new applications (Teubner et al., 2019). Investigating how trust and other determinants can be transferred from existing payment systems to cryptocurrencies could provide valuable insights for fostering user acceptance of these innovative technologies. The following research question arises:

- How can trust and trust-related determinants be transferred from existing systems to innovations (cryptocurrencies) and how do transferred attributes affect the intention to use these innovations?

The intricate nature of human behavior is shaped by a multitude of conscious and unconscious decisions (Evans, 2008), as well as a multitude of heuristics and biases (Tversky and Kahneman, 1981). The objective of nudges is to influence the underlying psychological effects as a strategic intervention, thereby rendering consumer decisions more predictable (Berger et al., 2022). The global fashion industry is one of the largest contributors to greenhouse gas emissions and environmental degradation (McKinsey & Company and Global Fashion Agenda, 2020). Digital nudges represent a potential solution in this regard, as they can subtly influence consumers towards more sustainable behaviors without the need for compulsion or sanctions (Berger et al., 2022). For instance, digital nudges that highlight the environmental benefits of eco-friendly fashion choices or remind consumers of the environmental challenges due to fast fashion could promote more sustainable purchasing behaviors (Tang et al., 2020).

Prior research has verified the efficacy of nudges in fostering pro-environmental behaviors, including recycling, energy conservation, and sustainable product choices (Berger et al., 2022; Maki et al., 2016). In the context of fashion retailing, the implementation of nudges can facilitate the advancement of sustainable behavior by prompting consumers to select environmentally conscious options or to reduce excessive consumption. This might contribute to the reduction of the industry's overall environmental impact (Tandon et al., 2023). In light of the growing consumer awareness regarding sustainability, nudges that resonate with this attitude may assist in bridging the gap between consumers' intentions and their actual purchasing behavior (Lee et al., 2020). While digital nudges offer considerable potential for promoting sustainable consumption, they also rise ethical concerns, particularly with regard to consumer autonomy, being manipulative and subtly influencing consumers to make choices that they may not have made independently (Schmidt and Engelen, 2020). This ethical conflict becomes especially salient when the objective of promoting sustainability is pursued through the use of nudges, a goal that many may consider to be desirable, but which may still encroach upon the autonomy of consumer decision-making. From an ethical perspective, the issue pertains to the extent of consumer awareness regarding the nudging process and the potential for their freedom to make unmanipulated choices to be limited. Consumers may reject the choice if they perceive that their autonomy is being eroded. Schmidt and Engelen (2020) posit that when digital nudges are perceived as overtly manipulative, they risk a negative response and the loss of trust in digital platforms. This presents an ethical dilemma: on the one hand, promoting sustainable behaviors aligns with societal goals to reduce environmental harm, but on the other hand, the use of digital nudges can be experienced as ethically questionable. Nevertheless, in comparison to bans or regulations, digital nudges represent a less intrusive approach to achieve pro-environmental behavior (Mirbabaie et al., 2021; Sunstein, 2014). They can subtly encourage sustainable actions without eliminating consumer choice, thus representing a compromise between complete consumer autonomy and stricter, top-down interventions. In light of the conflicting perspectives on the ethics of digital nudges and their potential for promoting sustainability, this study seeks to explore how these tools influence consumer behavior in online fashion retailing. In summary, the fourth research question is:

- How do digital nudges impact consumers' pro-environmental and purchase-usage related behavior in online fashion retailing?

The return of goods in e-commerce represents a relevant source of carbon emissions, largely due to the increased transportation and packaging waste that it entails (Asdecker and Karl, 2022). In fashion retailing, uncertainties about product fit represent a main reason for returns (Lee and Xu, 2020). Decision aids like tools for size advice can be used for supporting and optimizing consumers' size selection in online retailing (Abdulla and Borar, 2017; Nestler et al., 2021) and therefore lower the overall return rate, thus decreasing the environmental impact associated with product returns (Yang and Xiong, 2019). This is advisable due to overproduction and overconsumption in the fashion industry (Munir, 2020). While digital nudges have been the subject of criticism for potentially undermining consumer autonomy by subtly guiding decisions without explicit awareness (Schmidt and Engelen, 2020), decision aids present a more transparent alternative. Consumers actively engage with decision aids, making deliberate decisions based on the recommendations they receive (Yang and Xiong, 2019). This transparency ensures that consumer autonomy is preserved and the technology does not manipulate choices but rather provides users with additional information to make better decisions (Vugts et al., 2020). Furthermore, decision aids' ability to reduce returns aligns with sustainability goals without compromising transparency or consumer independence.

Despite the potential advantages of decision aids, they are not readily accepted by all consumers. The complexity of engaging with novel technologies can cause resistance, particularly if users perceive a loss of control to an algorithmic process (Ghasemaghaei et al., 2019). For example, users may ascribe unfavorable outcomes (such as receiving an incorrect size) to the technology, which can result in dissatisfaction (Moon and Nass, 1998). Furthermore, trust is a pivotal element, as consumers require assurance that their data is being utilized in an ethical manner (Wang and Benbasat, 2016). The adoption of decision aids may also be affected by perceived risks related to privacy and data usage of algorithmic recommendations (Nguyen and Khoa, 2019). Given the ethical benefits and sustainability potential of online size decision aids, it is crucial to explore the factors that influence consumers' intention to adopt and use these tools. This dissertation aims to investigate the following research question:

- What aspects influence consumers' intention to use decision aids for online size selection?

Considering the sustainability of innovative technologies such as electromobility, it is crucial to recognize the multifaceted nature of sustainability, as discussed by the Triple Bottom Line (TBL) framework proposed by Elkington (1997). The framework highlights that sustainability is not solely confined to environmental considerations but also encompasses economic and social sustainability (Purvis et al., 2019). By investigating the multidimensional nature of sustainability in the context of last-mile electrification, this comprehensive approach ensures that the transition to electromobility is not only advantageous for the environment but also economically viable and socially just.

Regarding environmental sustainability, the transition to electric vehicles in last-mile delivery offers significant promise in reducing carbon dioxide (CO_2) emissions and in mitigating climate change (Morganti and Browne, 2018). In addition to considerations related to emissions reduction, it is essential to incorporate factors such as the lifecycle impact of batteries and charging infrastructure to ensure long-term environmental sustainability (Tsakalidis et al., 2020). The economic dimension assesses the financial viability of electromobility for last-mile companies. While the initial investment in electric fleets and charging stations may be substantial, long-term operational savings from reduced fuel and maintenance costs could offset these expenses (Morganti and Browne, 2018). Furthermore, consumer demand for environmentally conscious deliveries may lead to a competitive advantage for companies that adopt electromobility, thereby contributing to economic sustainability (Silva et al., 2023). The social dimension of sustainability pertains to the well-being of employees and the broader community. For example, the transition to electric vehicles in logistics might improve work environment, but also include challenges such as range anxiety or delays in charging (Wolff and Madlener, 2019). It is imperative for companies to evaluate the broader implications of electromobility on urban environments, including noise reduction and enhanced air quality, which contribute to social well-being (Hardman et al., 2018).

Attitudes towards electromobility can differ between companies and employees. From a company perspective, the decision to adopt electromobility is influenced by factors such as regulatory incentives, cost-benefit analyses, and long-term operational considerations (Lebeau et al., 2016). For employees driving the electric vehicles, factors like ease of use, the reliability of the vehicles, and charging infrastructure can influence their acceptance or resistance to the technology (Juan et al., 2016). Considering the multifaceted nature of sustainability in last-mile electrification, it is imperative to comprehend the attitudes of both companies and employees towards electromobility. These stakeholders must reconcile the environmental, economic, and social dimensions of the transition, when determining whether to adopt electric vehicles. In summary, the sixth and final research question is:

- How do last-mile companies and their employees evaluate sustainability of last-mile electrification?

Structure and Content of the Essays 2

2.1 Focus of the Essays

The dissertation investigates consumers' and partly companies' behavioral intentions to use innovative technologies. Six essays aim to gain a better understanding of both drivers of acceptance and drivers of resistance of technology. These essays consider different technologies and innovative processes including AI, automated purchasing processes, cryptocurrencies, digital nudging, decision aids, and electromobility. Some of the essays follow a holistic approach, while others consider specific areas of application. Different methodological approaches have been chosen for the studies, including a systematic literature review, quantitative research, and qualitative research. Overall, implications for various stakeholders, such as companies, marketers, developers, or policymakers, can be derived from the dissertation.

Table 2.1 offers a short overview of the six essays, encompassing the research topic, the chosen method, results, and the technological context. In total, the six essays involve 1,467 participants in online surveys and interviews. In addition, 262 scientific articles were systematized.

J. Selter, *Acceptance and Resistance of Innovative Technologies*, Handel und Internationales Marketing Retailing and International Marketing,
https://doi.org/10.1007/978-3-658-48594-8_2

Table 2.1 Summary of the Six Essays

Essay	Topic	Method	Results	Technological Context
Essay 1: Ethics and Morality in AI—A Systematic Literature Review and Future Research	Investigates and systemizes the aspects of ethics and morality in the context of AI.	A systematic literature review was conducted including the years 2017 and 2021 resulting in 2,561 articles of which 262 articles were systemized.	Essay systemizes literature into seven topics and formulates future research questions for each topic.	AI
Essay 2: Aspects driving customers' intention to use automated purchasing processes	Investigates the usage intention of automated buying processes in a B2C retail context and highlights the importance of trust and behavioral control.	An online experiment (2×2 between subject design) was conducted with N = 176.	Results show decreasing user intention for automated shopping processes. Trust and behavioral control mediate the effect, which is further moderated by inherent novelty seeking.	Automated purchasing processes
Essay 3: Transferring Trust, Risk and Security on Users' Intention to Trade Cryptocurrency, Using PayPal as an Example	Examines the transfer of trust, perceived risk and security from the payment service provider PayPal to cryptocurrency.	An online survey (N = 398) was conducted in Germany and an explorative factor analysis was used.	Significant effects were found for the transfer of the examined attributes from PayPal to cryptocurrency if handled by PayPal and from cryptocurrency if handled by PayPal on the intention to use.	Cryptocurrencies

(continued)

Table 2.1 (continued)

Essay	Topic	Method	Results	Technological Context
Essay 4: Digital Nudging to Reduce Product Returns in E-Commerce: Investigating Consumers' Intentions and Environmental Implications	Explores the influence of default, framing, social norm, graphical, and incentive digital nudges for reducing wrong or doubled ordered sizes in online retailing.	Two online experiments with six scenarios (N = 352) and two scenarios (N = 246) were conducted.	Significant effects for the digital nudges on consumers' pro-environmental as well as purchase and usage related behavior were measured. Mediating effects of hedonic, normative, and gain goal frames could be shown.	Digital Nudging
Essay 5: Decision Aids for Online Size Selection: Insights from Experimental Studies in Apparel Retailing	Investigates decision aids for reducing wrong or doubled ordered sizes in online retailing and focuses on innovativeness of the aids and the degree of data disclosure required.	Two online experiments with two scenarios (N = 138) and three scenarios (N = 131) were conducted.	Both higher levels of data disclosure and innovativeness of the decision aid reduce usage intention. A higher level of data disclosure negatively influences consumers' intention to disclose data, and innovativeness has no effect on the intention to transact in general. Mediating effects for complexity and risk have been shown.	Decision Aids

(continued)

Table 2.1 (continued)

Essay	Topic	Method	Results	Technological Context
Essay 6: Sustainability Assessment of Last-Mile Electrification: A Qualitative Study in Germany	Explores the acceptance and attitude of courier, express and parcel service providers, subcontractors, and drivers regarding the sustainability of electric vehicles in the last mile.	A total of N = 26 in-depth interviews were conducted. Moreover, qualitative content analysis by structuring the content was conducted.	For the last-mile, economic sustainability plays a decisive role. However, this is not yet given according to the interview participants on electric light commercial vehicles.	Electromobil-ity

The theoretical background, method, findings, and implications of each essay are summarized in the following sections.

2.2 Essay 1—Ethics and Morality in AI—A Systematic Literature Review and Future Research

Theoretical Background

AI has become an important part of everyday life (Antao, 2019) and offers users several benefits with its increased processing power and the usability of big data (Huang and Rust, 2021). However, beside these advantages, ethical problems arise due to the development of AI and can occur as manipulation (Kaplan and Haenlein, 2019) or lost control (Steels and Lopez de Mantaras, 2018). Former research in this context has resulted in various research directions in the last years. Literature addresses ethical guidance (Vakkuri et al., 2020), different application areas (Hamilton and Davison, 2021), ethical governance (Shneiderman, 2020), or ethical aspects in programming (Arvan, 2018). However, a holistic overview of the topic is still lacking (Loureiro et al., 2021) and no paper could be identified, dealing with a systematization of the different research foci and an identification of the relevant articles in this context. Such a systematic review of the literature is important for identifying research gaps and directions for future research (Briner and Denyer, 2012). Here, the first essay of this dissertation tries to fill the gap in literature by answering the following research questions: (1) What is the current state of literature for ethics and morality in the context of AI? (2) What are future research areas in the context of ethics and morality with AI?

Method

To answer the research questions a systematic literature review was conducted following Briner and Denyer (2012) subsumed in five steps. After having defined the research questions, databases and Boolean Operators have been identified for the systematic review. ScienceDirect, Web of Science, JSTOR, EBSCO and Google Scholar have been searched and the publication period of the years from 2017 to 2021 was chosen, as the study was conducted at the beginning of 2022 and almost 90 percent of the articles relevant to the topic were published from 2019 to 2021 at that time. The search resulted in a total of 2,561 articles across all databases, considering all search criteria. In the following, these articles were screened for being found multiple times, being not peer-reviewed or just being a reproductive paper. Thus, 557 articles were selected for a review of the full text. Following several more exclusion criteria like impact factor, a lack of focus on ethics and morality or focusing on technical aspects, 262 articles meet the criteria and form the basis for the presentation of the current state of the scientific literature. To systematize the research articles, research methods, regions, results and focal points of the 262 contributions were protocolled and systematically analyzed for similarities and differences. It was found that systematization of the articles based on a consolidation of identical or similar research foci is best suited to provide an understandable overview of the current state of the literature. For a systematization reflecting the heterogeneity of the literature, seven main topics were identified.

Findings and Implications

Results show that disadvantages and concerns as well as advantages and opportunities of AI were addressed from an ethical and moral perspective. Since these corresponding articles depict immediate implications of the development and application of AI, they are considered in the first category. In the second category, findings of an ethical approach to AI that have been highlighted and discussed in recent years are explained, and in the third category, demands for governance and regulation of AI are reflected. In addition, requirements for programming that have been made in recent years were frequently addressed. The following category includes articles in which human moral judgement was explored more deeply in the discourse around AI. Other research foci resulting from the protocolling includes discussion of moral responsibility for consequences of an

AI decision or action and discussion of moral status of AI. Finally, articles that address law or policy against the backdrop of ethics, morality, and AI were identified. The essay provides a holistic overview of the literature of ethics and morality in AI and points out important research directions for further research. Moreover, practical implications can be derived, like the importance of close collaboration between science and policy to enable governance and regulation for an ethical approach to AI.

2.3 Essay 2—Aspects Driving Customers' Intention to Use Automated Purchasing Processes

Theoretical Background

With the use of AI, IoT devices have become increasingly autonomous (Willems, 2022) and smart devices represent an important part for private households and retailing (Gao and Bai, 2014). One example of smart devices is fully automated smart refrigerators, which can simplify the shopping process for consumers by automatically reordering groceries from retail stores (Hoyer et al., 2020). Automated purchasing processes in a B2C context are hardly considered in research (Marikyan et al., 2021) and the influence and impact of smart home technologies on the customer experience are still largely unexplored (Gonçalves et al., 2020). Even less amount of research could be identified for smart refrigerators, with only a few studies focusing on the consumer (Baier et al., 2016) or technical (Gupta et al., 2021b) perspective. Essay 3 consults the theory of technology readiness and the interplay of potential drivers and barriers of using new technologies that can be devided into different dimensions (Parasuraman, 2000). In addition to this, hedonic and utilitarian components can have different impacts on the preference or use of automated systems (Longoni and Cian, 2022). Thus, following research questions have been formulated: (1) Which role do hedonic and utilitarian utility components of products play regarding the intention to use an automated purchasing process? (2) How is the intention of using an automated purchasing process affected by trust, behavioral control and inherent novelty seeking?

Method

To answer these research questions an experimental study with a 2 (automation: not given vs given) $\times$ 2 (product: utilitarian vs hedonic) between-subject design was conducted. All scenarios were depicted as textual explanation and N = 176 ($M_{age} = 28.63$, SD = 11.51, 53.4% women) individuals participated. In the "non-automated" scenarios, participants were told to open their refrigerator and identify some missing groceries. Furthermore, they were told to procure these groceries in the typical way for them. In the "automated" scenarios, participants were told they had a smart refrigerator with cameras working with AI. The application would identify missing groceries on its own and reorder them automatically to their home based on previously adjusted settings. After being randomly assigned to one scenario, the respondents evaluated the automatization, the purchased products, and additional variables like trust, the feeling of control, and usage intention. To verify the hypotheses and to explore the mediation and moderation variables, ANOVAs were conducted using SPSS.

Results and Implications

Results of the second essay show a lower usage intention for automated shopping processes and no significant differences between utilitarian and hedonic product types. In addition, trust and behavioral control mediate and inherent novelty seeking moderate the effect of automation on usage intention. The second essay contributes to literature by extending literature regarding automated purchasing processes and IoT technologies from the consumer perspectives. Furthermore, this essay increases the amount of literature of consumer research regarding smart refrigerators being highly limited due to its new character and further extends literature regarding hedonic and utilitarian utility components in this context. Moreover, implications for practice can be derived. When intending to support automated shopping processes, retailers should adopt strategies to keep switching costs as low as possible and create an optimized customer experience.

2.4 Essay 3—Transferring Trust, Risk and Security on Users' Intention to Trade Cryptocurrency, Using PayPal as an Example

Theoretical Background

Former literature has shown different critical factors affecting cryptocurrency acceptance and use, including trust (Arli et al., 2021), risk (Mendoza-Tello et al., 2019), and security concerns (Rehman et al., 2020). Moreover, aspects such as trust, that are associated with an online service provider, can also be transferred to collaborating providers, companies, or services (Cao et al., 2018; Stewart, 2003; Wang et al., 2013). For instance, research has demonstrated that users are willing to transfer trust from a familiar to an unfamiliar website (Stewart, 2003). Based on this, this study is grounded in the Trust-Transfer Model (Stewart, 2003). Since the determinants of risk and security are also linked to the institution, they follow the same process of establishment of institution-based trust. Based on the literature and referring to the Trust-Transfer Model, essay 3 address the following research questions: (1) Is it possible to transfer perceived trust, risk and security in / of PayPal to cryptocurrencies when handled by PayPal? (2) Do these transferred attributes have an influence on the intention to a) use cryptocurrencies if handled by PayPal and b) use cryptocurrencies in general?

Method

To answer these research questions, essay 3 uses PayPal as a case study. Kalinic et al. (2019) have shown in their review that perceived trust, risk, and security are among the main drivers of customers' intention to use mobile payment systems like PayPal. A questionnaire was designed and distributed on an online platform and via social media platforms. Moreover, participants interested in finance were addressed via a larger social media channel, frequently dealing with financial topics. For maintaining a relevant and controlled sample, participants were questioned about their prior experience with PayPal and with cryptocurrencies. Additionally, the survey included a general explanation of cryptocurrencies. After data cleaning, the data set comprises $N = 534$ participants. To test the hypotheses, regression analyses were calculated using SPSS 27.

Results and Implications

Results illustrate that perceived trust in PayPal, perceived risk in PayPal, and perceived security of PayPal lead to an increased trust in, risk of, and security of using cryptocurrencies if they are handled by PayPal. The perceived trust in cryptocurrencies and perceived security of cryptocurrencies if handled by PayPal significantly increase the intention to cryptocurrencies if handled by PayPal and cryptocurrencies in general. Furthermore, perceived risk of using cryptocurrencies if handled by PayPal significantly decreases users' intention to use cryptocurrencies if handled by PayPal and cryptocurrencies in general. This essay contributes to literature by showing the transfer of several attributes between service provider and payment method. Additional insight into the ways in which attributes such as perceived trust, risk, and security can be transferred are provided, also justifying the consideration of extending the Trust-Transfer Model. These two extensions of the model provide new theoretical insights and broaden the understanding of IC-service usage. Moreover, added value for practical applications can ber derived. Different cryptocurrency providers, also including social media platform companies or state-owned central banks, can better understand the transfer of perceived attributes when working with payment providers.

2.5 Essay 4—Digital Nudging to Reduce Product Returns in E-Commerce: Investigating Consumers' Intentions and Environmental Implications

Theoretical Background

For guiding individuals decision-making process, (digital) nudges can be used modifying the choice architecture (Berger et al., 2022). Former literature has already investigated nudging elements to guide individuals in decision-making in physical (Dolan et al., 2012) and digital environments (Berger et al., 2020) and examined digital nudges as an effective approach changing consumers' pro-environmental behavior (Maki et al., 2016). However, research is lacking regarding the relevance of digital nudges in the context of e-commerce and hereby for reducing the number of returns. In addition, a research gap was identified regarding how digital nudges are influenced by different goal frames. According to goal framing theory, the hedonic goal frame, normative goal frame, and gain goal frame are particularly important in the context of pro-environmental behavior (Lindenberg and Steg, 2007). Each goal frame corresponds to a behavioral motivation (Tang et al., 2020). Based on the literature regarding digital nudges and goal framing theory, the following research question is addressed: Do digital nudges promote consumers' pro-environmental and purchase usage related behavior in mobile shopping app and how is the intention to place a selected order and to return influenced by consumers' goal motivation?

Method

To answer this research question, two online experiments have been conducted. First, an experimental study with six scenarios using a between-subject design was conducted. A fictional, app-based online store with the corresponding shopping cart as well as the information about having the same product in two different sizes in the shopping cart was presented to the participants. Based on Weinmann et al. (2016) five digital nudges (integrated default, social norm, graphical, incentive, and framing) were integrated. A total of 352 participants (M_{age} = 28.76, SD = 8.66) participated in the study and hypotheses were tested by calculating ANOVAs using SPSS. In the second study, an online experiment with two scenarios using a between-subject design was conducted to investigate the digital nudge framing and its impact on consumers' pro-environmental behavioral intention in more detail. To test the hypotheses and examine the mediating effects of the three gain goals, ANOVAs were calculated using SPSS and calculations with the SPSS add-on PROCESS were conducted.

Findings and Implications

Findings demonstrate that digital nudging effectively encourages pro-environmental behavior in fashion online shopping. Additionally, the essay reveals the critical role of hedonic, normative and gain goal frames in shaping pro-environmental attitudes. This essay contributes to literature by showing the effectiveness of default, social norm, graphical, incentive, and framing nudges in one study to influence consumers' behavioral intentions and further shedding light on the psychological mechanism underlying digital nudges. Hereby, goal framing theory was examined in this context (Lindenberg and Steg, 2007). Moreover, the essay contributes to literature by identifying the framing nudge to be the most promising method of influencing consumers' pro-environmental, purchase and usage related behavioral intentions and social norm to reducing consumers' intention to purchase to reuse the mobile app. Furthermore, important practical implications can be derived from the essay. digital nudges can be effectively integrated into mobile app's user interfaces, enhancing the decision making of their consumers by guiding and encouraging their behavior.

2.6 Essay 5—Decision Aids for Online Size Selection: Insights from Experimental Studies in Apparel Retailing

Theoretical Background

The approaches of digital nudges to reduce the number of returns presented in essay 4 might be ethically questionable as consumers are unconsciously pushed to a certain behavior. In contrast, decision aids can actively support consumers' decision making (Kalpoe, 2020). To effectively use decision aids, the disclosure of personal data often is required (ASOS, 2024; Zalando, 2023). Thus, essay 5 refers to privacy calculus theory (Dinev and Hart, 2006b; Laufer and Wolfe, 1977) and the weighing of risks and benefits of data disclosure (Culnan and Armstrong, 1999). Moreover, essay 5 refers to innovation resistance theory (Ram and Sheth, 1989) for investigating the influence of innovativeness of decision aids on consumers' usage intention. The the following research question was formulated: How is the intention to use decision aids for online size selection influenced a) by the degree of disclosed data, and b) by their innovativeness?

Method

To answer this research question, two online experiments have been conducted. In both experiments, a fictious online store with a shirt was depicted to participants and participants were asked to choose the size of the shirt. First, an experiment was conducted, using a between-subject design with two scenarios. For supporting decision making, each subject was presented one of two decision-making aids with different degrees of data input required. A total of 138 individuals (M_{age} = 27.4, SD = 9.76) participated in the study. In a second experiment, a between-subject design with three scenarios was conducted and innovativeness of the decision aids manipulated. A total of 131 participants (M_{age} = 25.7, SD = 6.73) participated in the study. Hypotheses of both studies were tested by using SPSS to calculated ANOVAs.

Findings and Implications
Results show that both higher levels of data disclosure and innovativeness of decision aids reduce consumers' usage intention. Furthermore, a higher level of data disclosure negativeley effects consumers' intention to disclose their data. No effect of innovativeness on consumers' intention to transact in general could be measured. Mediating effects of perceived complexity and perceived risk could be measured. This essay contributes to research by considering the disclosure of data when using decision aids as well as the innovativeness of these decision aids. Furthermore, research regarding complexity, risk and trust in privacy in the context of decision aid was extended. Additionally, practical implications can be derived from this essay. Retailers should enhance their decision aid portfolio step by step for not deviating too much from the current status-quo. In addition, the input of data required by consumers should be rather low and complemented by accurate information about security and transparency.

2.7 Essay 6—Sustainability Assessment of Last-Mile Electrification: A Qualitative Study in Germany

Theoretical Background
Former literature in the logistics context has pointed out the urgency of switching to alternative propulsion systems (Morganti and Browne, 2018) to mitigate pollution, combat global warming, and reduce reliance on fossil fuels (Liao et al., 2017). Several studies have shown that switching to ELCVs can be a promising way for reducing the environmental impact of the last mile (Iwan et al., 2021). However, when considering sustainability, not only the ecological perspective, but also the economic and social dimensions should be taken into account (Elkington and Rowlands, 1999). This multidimensionality of sustainability has already been used in the context of urban freight transport (Mello Bandeira et al., 2019),

highlighting the importance of investigating the TBL model in the context of sustainable last-mile logistics (Silva et al., 2023). Furthermore, previous research has also demonstrated the importance of considering different user groups and their varied expectations and attitudes in the context of electrification (Lebeau et al., 2016; Wolff and Madlener, 2019). However, most studies do not focus on comparing different user groups and research comparing the varying attitudes of CEP service providers, subcontractors, and drivers towards the sustainability of electromobility is lacking. Thus, essay 6 addresses the following research questions: (1) How do logistics service providers evaluate the use of electric vehicles in the last mile from environmental, economic, and social perspectives? (2) Which contrasting or concurrent perspectives can be identified in this context between CEP service providers, subcontractors, and drivers?

Method

To answer these research questions, a qualitative study with 26 interviews was conducted. A semi-structured guideline was developed (Mayer, 2013), based on previous research of Wikström et al. (2016) and Tsakalidis et al. (2020). The study took place in Germany and the interview questions focused on the experience and expectations of electric vehicles and charging infrastructure in the last mile. Moreover, the interviews included questions about current processes. For getting a holistic overview about the different levels included in the last-mile, managers of CEP service providers and subcontractors as well as drivers have been interviewed. The interviews were recorded, anonymously transcribed, coded, and subsequently analyzed using MAXQDA software. Based on Mayring and Fenzl (2019) approach to content analysis, the main categories were formed by a deductive and the subcategories were formed by an inductive approach.

Results and Implications

Results show that some statements cannot easily be assigned to one of the three main categories of ecological, economic, and social sustainability. Thus, a fourth category, technical process-related aspects, was added. According to the number of subcategories and coded text passages, the economic perspective was discussed most extensively. The interviews have shown that economic sustainability of last-mile electrification tends to be viewed negatively and plays the most important role considering the switch towards ELCVs. While technical process-related aspects were also considered negatively, ecological, and social sustainability were mentioned as positive side effects. The sixth essay contributes to literature, by showing the value of a multidimensional approach for evaluating sustainability of a technology. Furthermore, the added value of expanding ecological, economic, and social sustainability with technical process-related aspects when considering last-mile electrification was shown. Important implications for practice can be derived the attitudes of the different stakeholders—service providers, subcontractors, and drivers—should be examined for a successful electrification of the last-mile. Additionally, an exchange between the mentioned levels is important for last-mile companies and governments.

3 Essays

3.1 Essay 1—Ethics and Morality in AI—A Systematic Literature Review and Future Research

This study was published as: Selter, J. L.; Wagner, K. & Schramm-Klein, H., 2022. Ethics and Morality in AI—A Systematic Literature Review and Future Research. Proceedings of the 2022 ECIS, 60.

The essay in this dissertation is partly adapted compared to the published study.

3.1.1 Introduction

Applications from voice assistants and platforms such as Siri, Alexa, Netflix, and YouTube are based on AI and are increasingly finding their way into everyday life (Antao, 2019). AI has established itself in almost all application areas, such as construction (Pillai and Matus, 2021), telecommunications (Macnish and Fernandez Inguanzo, 2019), and automotive industries (Cunneen et al., 2020). AI can also be used to benefit people in the health care sector (Togni et al., 2021) or in the legal system (Simshaw, 2018) through its integration in both physical (Szocik and Abylkasymova, 2021) and digital (Smith and Miller, 2021) applications and technologies.

AI offers users a wide range of benefits through increased computing power at lower costs and usability of big data (Haenlein and Kaplan, 2019; Huang and Rust, 2021). For example, AI can analyze user emotions and prepare the script for an advertisement based on them (Huang and Rust, 2021), generate a personalized playlist on platforms such as Spotify based on user data (Puntoni et al., 2021),

J. Selter, *Acceptance and Resistance of Innovative Technologies*, Handel und Internationales Marketing Retailing and International Marketing,
https://doi.org/10.1007/978-3-658-48594-8_3

or automate the clothing buy-back process through a chatbot (Schanke et al., 2021). However, not only benefits but also ethical problems have accompanied the development of AI, such as the prioritization of and discrimination against certain users based on demographic and economic aspects of customer relationship management (Libai et al., 2020) or the unequal representation of providers and companies on platforms such as Airbnb or Amazon, leading to unequal market shares (Milano et al., 2021). Additionally, with the support of AI, attempts have already been made to manipulate people through targeted political advertisements based on social media data (Kaplan and Haenlein, 2019) and to monitor and log user behavior (Du, 2021). Entrepreneurs such as Elon Musk and Bill Gates also see risks in using AI, including losing control over its use (Steels and Lopez de Mantaras, 2018). Here, autonomous weapons systems and associated issues regarding human safety (Hynek and Solovyeva, 2021) as well as ethical concerns with surveillance technologies (Belk, 2021) have been mentioned. Addressing the ethical implications of the application of AI is important because of the increasingly consequential interaction between humans and AI (McManus and Rutchick, 2019). The question of creating ethically designed AI-based solutions (Morley et al., 2020) arises from these controversial topics. The increasing emergence of ethics-related projects in technology associations (Adamson et al., 2019) and the development of ethical guidelines and principles by, for example, the EU Commission (EU Commission, 2020) or by China (Wu et al., 2020a) also highlight the international awareness of the relevance of ethics and morality in the context of AI development.

The development of AI and the accompanying public interest in it have resulted in various research directions over the last years. However, research has only provided insights into individual research areas in the context of AI and ethics, while a holistic overview of the topic is still lacking (Loureiro et al., 2021). The literature addresses, for example, ethical guidance (Vakkuri et al., 2020), different application areas (Hamilton and Davison, 2021), ethical governance (Shneiderman, 2020), and ethical aspects of programming (Arvan, 2018). However, no study has dealt with systematization of the different research foci and identification of the relevant articles in the context of AI and ethics. Such a systematic review of the literature is necessary to identify research gaps and new research opportunities (Briner and Denyer, 2012) and results are valuable for academia and practitioners that are involved in development, application, or commercialization of AI-based products and services. Thus, this research aims to analyze literature regarding the ethical and moral aspects in the context of AI to (i) identify and systemize the existing literature, (ii) provide an overview of the discussed topics in this context, and (iii) highlight future research directions and specific research questions. Therefore, our research questions are as follows:

Research question 1:	What is the current state of the literature on ethics and morality in the context of AI?
Research question 2:	What are the future research areas regarding ethics and morality in the context of AI?

3.1.2 Theoretical Foundation

This section provides a brief introduction to the relevant concepts of this study and an overview of the keywords used to create the theoretical foundation for the systematic literature review. Thus, a short definition of AI, including machine learning (ML) and deep learning (DL), is provided. Further, the context of morality, applied ethics, and machine ethics, which play a central role in the context of AI, is introduced.

Because AI has been applied in several contexts, different definitions of it are used (Wang, 2019). For example, an early definition of AI indicates that the term refers to the action of a computer when that action, if performed by a human, requires intelligent thought. This implies that intellectual capability is not limited to humans or other living beings but also exists among computers (Simon, 1995). The term "AI" is also used to describe multidisciplinary research in the fields of computer science, systematic neuroscience, and human cognitive science (Cichocki and Kuleshov, 2021). In general, AI can be subdivided into ML and DL (Fig. 3.1). ML, as one part of AI, represents the ability of systems to learn and improve automatically and independently of humans as long as the data are structured (Campbell et al., 2020). DL as another part of AI and a subarea of ML, is based on neural networks, computational networks that are biologically inspired by and operate based on the functionality of the human brain (Haenlein and Kaplan, 2019). DL includes computer vision, robotics and artificial agents, natural language processing, speech recognition and production, augmented and virtual realities, data analytics, and automatic planning, sequencing, and optimization (Cichocki and Kuleshov, 2021). Furthermore, AI can be divided into different degrees. While weak AI can outperform humans at specific tasks, next- generation AI can outperform humans at almost all cognitive tasks (Lu et al., 2018).

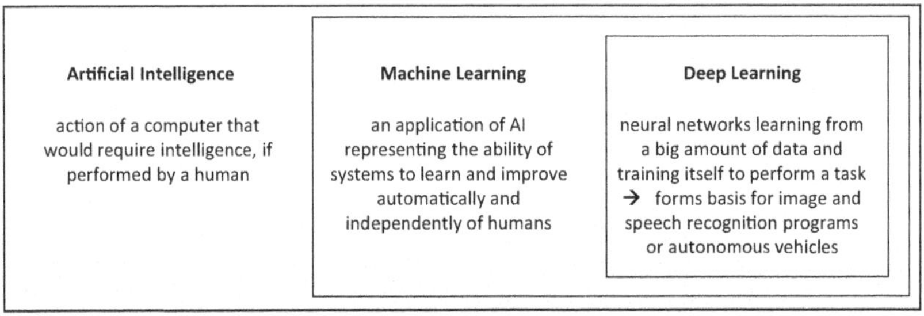

Fig. 3.1 Concept of Artificial Intelligence

The second part of this research, the concept of ethics and morality, encompasses the field of applied ethics. Applied ethics includes machine ethics (Bauer, 2020), which is of particular importance in the discourse on AI. It can be divided into two research directions. One direction includes robot ethics or AI ethics in the context of the avoidance of harm through the application of robots and AI. This research direction encompasses the behavior of developers, manufacturers, and users. The other direction includes the terms "machine ethics," "ethical AI," and "ethical robots" and refers to research on ethical behavior of robots and AI, as well as research on the implementation of ethical values in autonomous systems. Machine ethics was the first research field to deal with artificial morality (Winfield et al., 2019). Artificial morality explores the possibilities of the moral capabilities of artificial systems (Misselhorn, 2018). Therefore, the discourse addresses the moral status of AI, including moral theories directed toward moral patients and moral agents. Moral patients can be positively or be negatively affected by good or bad actions. Moral agents, meanwhile, can perform right or wrong actions (Danaher, 2021). Moral agency, and thus the ability to perform right or wrong actions, requires the perception and processing of information as well as the subsequent ability to act and learn (Danaher, 2019). This shows that the concept of morality, as distinct from the concept of ethics, is mostly directly related to decisions and actions (Verbeek, 2008). In the context of ethics, the question of what is a "right" action is considered (Fritz et al., 2019), whereas in the context of morality, the rules, values, and norms that influence a person's behavior are considered (Bartneck et al., 2019). Nevertheless, in the scientific discourse, a clear conceptual separation between ethics and morality does not always exist. The adjectives "ethical" and "moral" are used as synonyms in contrast to fixed terms such as "AI ethics" and "moral agency". For instance, the expressions "ethical machines" (Tolmeijer et al., 2020) and "moral machines" (Magrani, 2019) correspond analogously.

3.1.3 Methodology

The systematic literature review was conducted in accordance with Briner and Denyer (2012) and had five steps. After identifying the research questions and search strings, search and selection criteria were defined for the selected databases and the inclusion and exclusion protocols were set, including the publication period and type of articles. The review and synthesis as well as the report followed (Fig. 3.2).

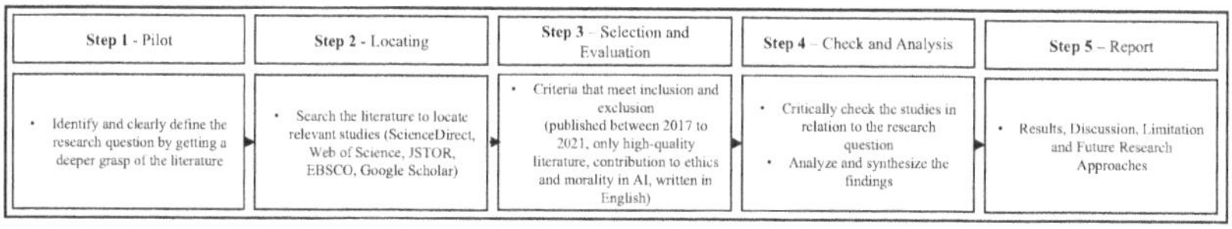

Fig. 3.2 Research Process of Systematic Literature Review

In the pilot step, we searched the literature to get a better understanding for the examined field and the existing literature. In this step, we formulated the research questions and identified the search strings (Table 3.1). At the same time, we identified inclusion and exclusion criteria for the literature. Because a well-formulated research question is crucial for a systematic literature review, the pilot stage led us to two research questions. During step 2, different search engines were selected for broad access to diverse relevant literature. Five databases were selected with a large amount of peer-reviewed literature related to our research question (ScienceDirect, Web of Science, JSTOR, EBSCO, and Google Scholar). Every database was searched using specified search strings (Table 3.1). These strings were the same for every database except for some minor adjustments.

Table 3.1 Search Functions and Results of the Selection of Research Articles

Database	Search String	Results
ScienceDirect	Title, abstract or author-specified keywords: artificial intelligence OR ai; Title: moral OR morality OR morals OR morally OR ethic OR ethically OR ethics OR ethical	103
	Title, abstract or author-specified keywords: moral OR morality OR morals OR morally OR ethic OR ethically OR ethics OR ethical; Title: artificial intelligence OR ai	170
Web of Science	moral* OR ethic* (Title) and ai OR artificial intelligence (Topic) or ai OR artificial intelligence (Title) and moral* OR ethic* (Topic)	1,283
JSTOR	(ti:(moral* OR ethic*) AND (ai OR artificial intelligence)) AND la:(eng OR en)	251
	(ti:(ai OR artificial intelligence) AND (moral* OR ethic*)) AND la:(eng OR en)	135
EBSCO	TI (ai or artificial intelligence) AND AB (ethic* or moral*) OR TI (ethic* or moral*) AND AB (ai or artificial intelligence)	369
Google Scholar	allintitle: ai moral OR ethic OR morally OR morality OR morals OR ethics OR ethical OR ethically	125
	allintitle: artificial intelligence moral OR ethic OR morally OR morality OR morals OR ethics OR ethical OR ethically	125

Regarding the inclusion and exclusion criteria, review articles, research articles, and mini reviews were searched for in Science Direct; proceeding papers, review articles, and early-access papers were searched for in Web of Science; journal articles were searched for in JSTOR; and scientific journals were searched for in EBSCO. Moreover, search in Google Scholar was conducted according to the relevance and finally reduced for both search algorithms to 125 results each, since no more relevant articles were found on the following results pages meeting all the selection criteria. Overall, the search resulted in a total of 2,561 findings across all databases, taking into account all search criteria and being published between 2017 to 2021. These 2,561 findings also include articles that were found several times. A two-stage selection process followed for the inclusion or exclusion of articles in the literature review. Articles that were found several times, were not peer reviewed, and did not contribute to answering the research questions on the basis of the title and abstract were excluded. Additionally, duplicate articles, book reviews, chapters, case reports, discussions, editorials, and news articles were excluded. Purely reproductive papers with no contributions of their own were also excluded. Thus, 557 articles were selected for a review of their full texts. Next, the impact factor, a quality indictor of scientific journals, was examined (Briner and Denyer, 2012). Articles from journals with a two-year impact factor below 0.5 were excluded. Therefore, of the 557 articles, a further 80 were excluded. During the full-text review, care was taken to ensure that the papers considered aspects of ethics and/or morality as well as AI, and especially to ensure that both aspects were considered in relation to each other. Thus, 126 articles with a lack of focus on ethics and morality, five articles with insufficient focus on AI, and 26 articles with no linkage among the topics were eliminated. Finally, 58 technical articles that did not provide concrete research results were eliminated. In the end, 262 articles met the criteria and formed the basis for reviewing the current state of the scientific literature on the topic of ethics and morality in the context of AI. Two reviewers read each paper to ensure it met quality requirements and to combat potential bias. Moreover, we defined the protocols for reviewing titles, keywords, and abstracts. For example, all articles had to be written in English, AI together with ethics and morality had to be the main focus, and the paper had to contribute to this field. We checked about 100 abstracts for intercoder reliability for the selected criteria. The abstracts were discussed, checked against the set criteria, and compared. Any disagreements were solved (Miles and Hubermann, 1994). Fig. 3.3 shows the inclusion and exclusion protocol.

According to Petersen et al. (2015) and Wohlin et al. (2012), there are different threats to validity such as construct validity, external validity, internal validity, and conclusion validity. To address these threats, we strictly followed the five steps Fig. 3.2 shows and adhered to the protocols for reviewing the articles. For generalizability, we not only followed the five steps but also described the research process. No statistical cause relationship was established, and we combated internal validity through regular meetings of the authors to combat potential bias. To meet the expectations for conclusion validity, two authors were involved, all 2,561 studies were regarded from the beginning of the review process until the 262 studies were identified. All reviewers were involved in ascertaining the findings and preparing the conclusion and discussion sections.

The fourth step involved double-checking the relevant literature as well as the analysis and synthesis. For the review, the papers were broken down to their individual components and their relationships were described. The synthesis involved examining the connection among the different papers. It involved inter-synthesis and intra-analysis (Tranfield et al., 2003). To establish a system for the analysis, the full texts of the 262 articles were rescreened. For each article, research methods, regions, results, and foci, as well as application areas, AI technologies, and aspects of ethics and morality were examined and listed. It was determined that systematization of the articles based on an amalgamation of identical or similar research foci was best suited to obtain an understanding and overview of the current state of the literature and to answer the underlying research questions. We were able to derivate the categorization of the different topics based on our research questions.

Step 5 involved reporting the results by showing the categorization of the different topics based on our research questions, answering and discussing the research questions based on the results, and indicating future research possibilities as well as limitations.

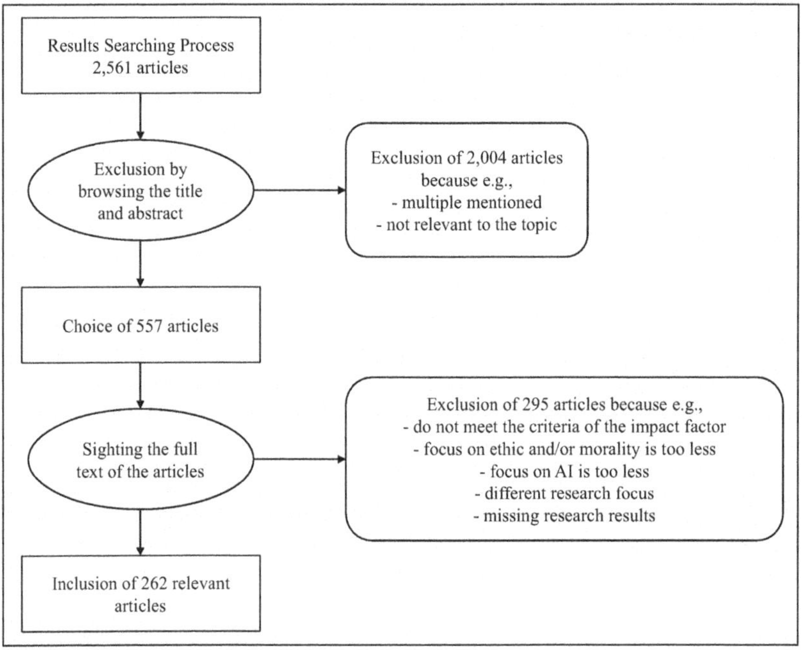

Fig. 3.3 Overview of the Inclusion and Exclusion Protocols

The selected articles were published in a total of 127 different journals. to provide an overview of the dominant journals, journals in which more than three of the selected articles were published are listed herein: AI & SOCIETY (36), Ethics and Information Technology (19), Science and Engineering Ethics (15), Minds and Machines (13), Big Data & Society (6), The ORBIT Journal (6), Computer Law & Security Review (5), Frontiers in Robotics and AI (5), International Journal of Social Robotics (4), IEEE Technology and Society Magazine (4), and AI Magazine (4). Furthermore, Table 3.2 offers an overview of all the journals whose articles were selected for this literature review.

Table 3.2 List of Journals and Number of Cited Papers in Each Journal

Number of Papers in the Journal	Journal
36	AI & SOCIETY
19	Ethics and Information Technology
15	Science and Engineering Ethics
13	Minds and Machines
6	Big Data & Society, The ORBIT Journal
5	Computer Law & Security Review, Frontiers in Robotics and AI
4	AI magazine, IEEE Technology and Society Magazine, International Journal of Social Robotics, Information Communication & Society
3	Procedia Computer Science, Cambridge Quarterly of Healthcare Ethics, Applied Artificial Intelligence
2	Journal of Business Research, Computers in Human Behavior, Social Science & Medicine, Telecommunications Policy, IFAC Papers Online, Futures, IEEE Access, Journal of Medical Ethics, Journal of Business Ethics, Nature Machine Intelligence, International Journal of Law and Information Technology, Artificial Intelligence, IEEE Security & Privacy, Scientific Reports, Journal of Consumer Research, Bull World Health Organ, Artificial Intelligence in Medicine, International Journal of Human Computer Studies, Philosophy & Technology, International Journal of Artificial Intelligence in Education, Journal of Information Communication and Ethics in Society

(continued)

Table 3.2 (continued)

Number of Papers in the Journal	Journal
1	Journal of Responsible Technology, Engineering, Canadian Association of Radiologists Journal, iScience, New Ideas in Psychology, Clinical Biochemistry, Diagnostic and Interventional Imaging, Cognitive Systems Research, Future Internet, European Journal of Operational Research, International Journal of Child-Computer Interaction, Technology in Society, Government Information Quarterly, American Journal of Opthalmology, IEEE Software, Critical Social Policy, Philosophical Transactions A, Diagnostics, Business and Human Rights Journal, New Media and Society, Society, Asian Journal of Law and Society, Cureus, Nature, Nature Human Behavior, Journal of Gastroenterology and Hepatology, Information & Management, Nursing Philosophy, Science Editing, London Review of Education, Neuroethics, Behaviour & Information Technology, Journal of Medical Internet Research, Contemporary Social Science, Bioethics, Frontiers in Cardiovascular Medicine, ACM Transaction on Interactive Intelligent Systems, Frontiers in Psychiatry, BMC Medical Ethics, Columbia Law Review, Journal of Clinical Psychology, Employee Responsibility and Rights Journal, Biomedical Informatics Insights, Nanoethics, IBM Journal of Research and Development, Australasian Journal of Information Systems, Australasian Journal of Dermatology, Life Sciences, Society and Policy, Artificial Intelligence Law, Public Management Review, Public Integrity, Future Internet, Medicine Healthcare and Philosophy, Radiology, Science and Public Policy, Public Integrity, Synthese, International Journal of Human Computer Interaction, Journal of the American Medical Informatics Association, Journal of Oral Biology and Craniofacial Research, Global Policy, Education and Information Technologies, AI Communications, Information System Frontiers, Psychology & Marketing, Sustainable Development, Journal of Communication, Proceedings of the 2019 AAAI/ ACM Conference on AI ethics and Society, Intensive Care Medicine, Journal of Artificial Intelligence Research, Journal of Dental Research, Computers in Biology and Medicine, Science & Education, The Journal of Systems & Software, SSM—Mental Health, The American Journal of Pathology, The Southern Journal of Philosophy, Journal of Medical Imaging and Radiation Oncology, European Business Organization Law Review, International Journal of Information Management, IEEE Robotics & Automation Magazine, Applied Sciences

Almost 90 percent of the relevant literature was published during the chosen time period. This is probably because AI set off major trends, and many AI-based applications emerged during this time. Moreover, within the Web of Science database, the number of published papers on ethics and morality in the context of AI was twice as high in 2017 than in 2016. AI is a fast-growing sector and a fast-developing technology, so it is important to include the latest trends. For example, voice recognition has decisively improved over the last few years, so studies on the topic from 10 years ago are not very relevant for today's research. However, our study aimed to indicate the current state of the literature of ethics and morality in the context of AI (research question 1). Furthermore, we were able to show, that most of the chosen 262 studies were published between 2020 and 2021 (Fig. 3.4). The distribution of the publication periods of the selected articles illustrates their high relevance to current events. Figure 3.4 shows that the number of publications on the topic of ethics and morality in the context of AI has steadily increased from 2017 to 2021.

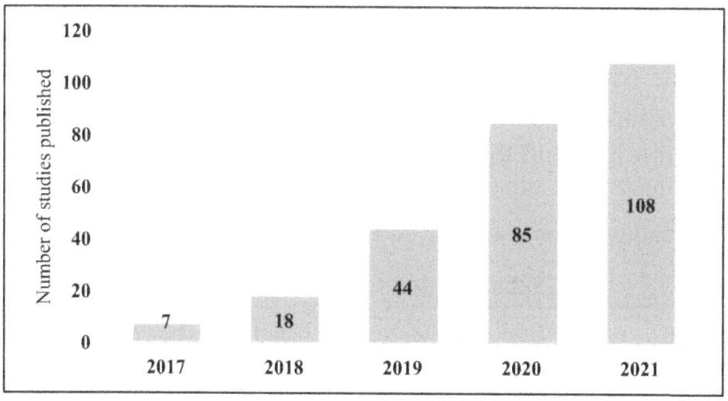

Fig. 3.4 Assignment of the 262 Selected Articles to the Years of Publication

3.1.4 Review and Results

During the review of the literature, it became apparent that the topic of ethics and morality in the context of AI has been discussed from a wide variety of perspectives and in different application areas. For a suitable systematization that reflected this heterogeneity, seven main topics were identified, which in turn were divided into different subtopics (Table 3.3). The exclusion and inclusion protocols indicated that disadvantages and concerns as well as advantages and opportunities of AI were addressed from an ethical and moral perspective. Because these articles discussed the immediate implications of the development and application of AI in different application areas as well as in the context of different applications and technologies, they were considered in the "Ethical and moral implications of AI" chapter. Next, findings of an ethical approach to AI that have been highlighted and discussed in recent years were discussed. Recommendations and principles for dealing with AI and further approaches were listed in the "Approaches to Ethical Handling of AI" chapter. The third chapter, "Demand for Governance and Regulations" included articles dealing with the integration of ethics in AI on a political level. Requirements for programming that have been devised in recent years were frequently addressed and mentioned in the "AI Programming Requirements" chapter. The next chapter, "Human Moral Judgement and Moral Responsibility," included articles in which humanity's capacity for moral judgment was explored more deeply in the discourse around AI. It also included discussions about moral responsibility of different stakeholders. Several articles discussed the moral status of AI today and in the future and were presented in the "Moral Status of AI" chapter. Finally, articles that addressed law or policy against the backdrop of ethics, morality, and AI were identified and summarized in the "Importance of Law and Politics" chapter. Table 3.3 illustrates these seven main topics and their subtopics.

Table 3.3 Systematization of Literature on Ethics and Morality in AI

Topic	Subtopics and Content
Ethical and Moral Implications of AI	Ethical concerns and benefits in the context of AI
	Ethical and moral implications in the context of AI in different application areas (business, health sector, public sector)
	Implications of physical products and digital products, and implications without specific application areas, uses or technologies
	Implications of the relationship of humans to machines
Approaches to Ethical Handling of AI	Recommendations about the use of ethics in the context of AI
	Principles of dealing with AI Further approaches and human rights

(continued)

Table 3.3 (continued)

Topic	Subtopics and Content
Demand for Governance and Regulations	Models and principles to implement existing principles as a bias for ethical governance
	Governance of AI on a political level
	Need for regulation of AI
AI Programming Requirements	Approaches to moral and moral justification as well as user and stakeholder values to be programmed into AI
	Programming of friendly AI that can recognize moral actions
	Integration of rebel agents
	Further technical and social challenges
Human Moral Judgement and Moral Responsibility	Moral judgement of humans and human moral values
	Moral responsibility of developers, providers and users
	Moral responsibility of AI itself
Moral Status of AI	Discussion about the moral status of AI
	Moral status of AI in the future
Importance of Law and Politics	Importance of the legal system to enforce regulations
	Legislative gaps and policies of various countries around the world

3.1.4.1 Ethical Implications of AI

The systematic review of the literature showed that many articles deal with the direct implications of the development and application of AI and specifically with the advantages and disadvantages as well as the problems with and opportunities of AI from an ethical and moral perspective. Ethical concerns addressed include invasion of privacy, unauthorized data collection (Truby and Brown, 2021), hacking (Swarte et al., 2019) concerns for humanity (Mao and Shi-Kupfer, 2021), and disadvantages to social groups due to biased data and a lack of transparency of AI (Pillai and Matus, 2021). The literature is also concerned with ethics-related benefits of the use of AI, such as satisfaction of customer needs through AI use in marketing (Hermann, 2021b), potential effectiveness of using AI to prevent suicide (Luk et al., 2021), and stress reduction associated with replacing human soldiers with autonomous unmanned aircrafts (Swarte et al., 2019).

Furthermore, the systematization of the literature showed that three application areas involve ethical and moral implications of AI. These are implications for businesses, for example, for back-end or on-site tasks in the construction sector or for supportive tasks in HR management (Colombino et al., 2021; Hamilton and Davison, 2021; Hermann, 2021a, 2021b; Macnish and Fernandez Inguanzo, 2019; Pillai and Matus, 2021; Ryan, 2019), implications for the health sector, for example, for comparing computer tomography images in radiology (Brady and Neri, 2020; Brown and Halpern, 2021; Fenech and Buston, 2020; Fiske et al., 2019; Goirand et al., 2021; Kempt and Nagel, 2022; Kerasidou, 2021; Kretzschmar et al., 2019; Luk et al., 2021; Martinho et al., 2021a; Montemayor et al., 2021; Mörch et al., 2021; Morley et al., 2021a; Park et al., 2019; Saheb et al., 2021; Sand et al., 2021; SFR-IA Group and CERF, 2018; Shaw et al., 2021; Stewart et al., 2021; Stokes and Amitabha Palmer, 2020), and public sector, for example, for understanding complaints of citizens or using chatbots for parking permits, and the education sector (Calvo, 2020; Chounta et al., 2021; Liu et al., 2019b; Mark, 2019; Ryan, 2019; Ryan and Anya, 2019; Sun and Ye, 2021; Swarte et al., 2019; Szocik and Abylkasymova, 2021; Vallejos et al., 2017; Wakunuma et al., 2020).

Next, the implications of specific applications or technologies, divided into physical and digital products, are discussed in the literature. Themed physical products include caregiver robots (Yew, 2021) and robots for police applications (Szocik and Abylkasymova, 2021), autonomous vehicles (Cunneen et al., 2019a; Reed et al., 2021; Siegel and Pappas, 2021; Vrščaj et al., 2020) and devices or toys for children (Antle and Kitson, 2021; McStay and Rosner, 2021). Digital applications include online surveillance (Nigam et al., 2021) and biometric facial recognition systems (Smith and Miller, 2021), recommender systems (King, 2020; Milano et al., 2020), digital twins (Popa et al., 2021; Truby and Brown, 2021), intelligent assistance technologies (Arifin and Lennerfors, 2022; Wangmo et al., 2019), and risk assessment tools (Liu et al., 2019b).

Some of the articles we considered directly deal with AI's implications without referring to application areas or technologies. They discuss the impact of AI on several aspects such as the environment (Murdock, 2018), human rights (Livingston and Risse, 2019), and human moral status (Danaher, 2019), and they explore ethical concerns and benefits of using AI. One study found that AI raises major concerns about an increase in unemployment and inequality (Ghotbi et al., 2021). Further, the decision-making power of AI can change the moral norms of humans (Gill, 2020). Whereas one study raised doubts (Novak, 2020), another suggested that AI can promote unethical human behavior (Köbis et al., 2021). In contrast, AI can train moral cognitive abilities and influence one's moral actions, thus promoting moral decision-making (Lara and Deckers, 2020). Moreover, the possible collaboration of AI with humans has been demonstrated (Seeber et al., 2018), with AI being able to change the relationship of humans to machines at the affective, relational, and physical levels as well as to expand the physical, mental, and physical scope of humans (Togni et al., 2021). Affective relationships between humans and AI have been further affirmed with regard to emotional AI. The importance of being aware of differences among human relationships has been emphasized (Weber-Guskar, 2021).

3.1.4.2 Approaches to Ethical Handling of AI

Our review of the literature indicated that it deals with different research directions, such as recommendations or principles for handling ethical and moral aspects in the context of AI. For example, Floridi et al. (2018) suggested recommendations for European policy makers called "Recommendations for a Good AI Society". Other publications have formulated recommendations explicitly for companies on how to consider ethical aspects when developing AI (Vakkuri et al., 2020) and how to solve ethical problems and develop corporate social responsibility (Du and Xie, 2021). Few authors have focused their research entirely on making recommendations for specific application contexts of AI (Asaro, 2019; Jotterand and Bosco, 2020; McKernan et al., 2018). For the most part, authors have made recommendations for specific application contexts (Brady and Neri, 2020; Calvo, 2020; Fiske et al., 2019; Hamilton and Davison, 2021; Jobson et al., 2022; Kretzschmar et al., 2019; Yew, 2021).

Recommendations can be distinguished from established principles in the discourse about AI. Principles for dealing with AI are most often mentioned in papers addressing AI applications in the health sector (Abràmoff et al., 2020; Chauhan and Gullapalli, 2021; Joerin et al., 2020; Kenny et al., 2021; Larson et al., 2020; SFR-IA Group and CERF, 2018; Stewart et al., 2021). Another context in which researchers have considered principles for dealing with AI is national defense (Taddeo et al., 2021). Other studies have generally considered how companies are applying these principles in practice (Ibáñez and Olmeda, 2021; van Bruxvoort and van Keulen, 2021). Principles for the entire context of AI ethics have also been established in recent years. For example, Floridi et al. (2018) brought together existing principles and identified the following main principles for ethical treatment of AI: beneficence, harm avoidance, autonomy, justice, and explicability. These principles have also been examined in follow-up research (Floridi et al., 2021) and discussed by other authors (Milossi et al., 2021). Scholars have also criticized these principles, such as their contradictions or dependencies (Hermann, 2021a; Robbins, 2019). Similarly, other recent studies have pointed out the limitations of using principles in the context of AI ethics (Mittelstadt, 2019; Whittlestone et al., 2019) and the challenges of implementing these principles (Peters et al., 2020; Stix, 2021), as well as the contradiction of these principles with the concept of ethics itself (Rességuier and Rodrigues, 2020). Although these principles have global similarities, the approaches to and motivations for implementing them are heterogeneous (Jobin et al., 2019; Morley et al., 2021b). The principles are not binding (Buruk et al., 2020) but merely a response to public demands (Kerr et al., 2020) without concrete contextual reference (Hagendorff, 2020). The literature has criticized models regarding the applicability of the principles (Bærøe et al., 2020) and shown that the principles need to be adapted to organizations' needs (Vakkuri et al., 2020) and stakeholders' regional and cultural ethical preferences (Gabriel, 2020; Segun, 2021a). In recent years, efforts have been made in academia to help organizations understand guiding principles and consider ethical issues with AI (Clarke, 2019; Ryan and Stahl, 2021).

Other approaches to integrating ethical aspects of different AI applications have been discussed in the literature. For example, Sekiguchi and Hori (2020) developed a tool for engineers to support implementation of ethical technological designs through comprehensibility and standardization. They showed that this can promote engineers' creativity and motivation for long-term projects. Moreover, researchers have pointed out the importance of developing new approaches to considering the organizational dimension when operationalizing ethics in the context of AI (Krijger, 2021). Gerdes (2022) discussed an approach to designing AI ethics, and Vakkuri et al. (2021) illustrated a method for implementing AI ethics. Furthermore, the literature has explicitly considered how users can be protected and how technologies can be used in a socially responsible ways (Lannon et al., 2021), for example, countering AI bias using technical methods (Tomalin et al., 2021), ensuring compliance with ethical principles in the context of social networks, and countering the spread of fake news (Salem et al., 2020). Kelley and Atreides (2020) developed an assessment protocol that takes into account the specificity of the emotions and emotional structures of AI. In contrast to approaches targeting engineers and researchers already working in this field, one paper targeted teachers of AI ethics and conducted case studies to help students understand the ethical and moral aspects of AI (Burton et al., 2017). Other research, rather than exclusively addressing researchers or developers of AI, has addressed users of AI. For example, an ethical model has been created as a thinking note approach to support AI development and implementation (Schrader and Ghosh, 2018). Ethics-based auditing through a third party has been used to check the behavior of AI for compliance with standards and principles (Mökander and Axente, 2021; Mökander and Floridi, 2021). Furthermore, models for AI-related management decisions (Bankins, 2021; Brendel et al., 2021), business analysis (Vidgen et al., 2020), health care (Vallès-Peris and Domènech, 2021), and ethical decision making (Anshari et al., 2021) have been developed. One model already in use in the field is the "Data Ethics Decision Aid" model, which focuses on the ethical use of data (Franzke et al., 2021). Further approaches have been developed to fundamentally sensitize stakeholders to ethical considerations, for example, through an ethical checklist in the context of mental health (Mörch et al., 2020) and through a questionnaire in radiology (Geis et al., 2019), cyber-physical-human systems (Khargonekar and Sampath, 2020), and the military (Wasilow and Thorpe, 2019).

A few papers have based the conception, development, and use of AI on human rights principles rather than ethics. These papers justify their focus on human rights by arguing that laws have been formulated worldwide on human rights, so a uniform level for regulating AI already exists (Gibbons, 2021). Moreover, this approach considers accountability and power asymmetries and is complementary to ethics (Fukuda-Parr and Gibbons, 2021). In this context, Mantelero (2018) has developed a management tool to assess AI development and its impact. This tool has already been tested in companies and could support the regulation of AI (Mantelero and Esposito, 2021).

3.1.4.3 Demand for Governance and Regulation of AI

Another key theme identified in this study is that ethical models and values are difficult to understand and implement in the complex contexts of technologies and therefore, governance and regulations are required, specifically, social and organizational governance mechanisms supported by external actors (Raab, 2020). For example, Shneiderman (2020) and Winfield and Jirotka (2018) have developed models to implement existing principles as a basis for ethical governance. Moreover, governance models are needed in relation to not only AI development, but also AI application (Lepri et al., 2021). This is evident from the development of governance models for AI use in healthcare (Reddy et al., 2020) and in the formulation of principles to support data governance (Janssen et al., 2020). Additionally, governance of AI also occurs at a political level (Radu, 2021; Wu et al., 2020b), so the role of government in the governance process should be considered alongside the scientific community (Almeida et al., 2021).

Although governance has been called upon to solve implementation problems of other models and tools, a variety of models can be observed (Liu and Maas, 2021; Stahl et al., 2021a). For governance development, organizations should first understand their responsibility for ethical behavior (Abrams et al., 2019), and multiple stakeholders should be involved in the development process (Floridi, 2018; Liu et al., 2019b). Complementary governance development requirements exist in terms of political governance and of the need for international cooperation among governments. There have been calls for political governance not to be dominated by the United States, Europe, and China but all societies affected by AI (ÓhÉigeartaigh et al., 2020).

Even though there is a link between governance and regulation, the former does not necessarily go hand in hand with the latter (Winfield and Jirotka, 2018), so developed models are not legally binding (Stahl et al., 2021b). There is an increasing emphasis on the need for regulation of AI (Almeida et al., 2021; Fernandes et al., 2020), on criticism of the difficulties in implementing principles (Iphofen and Kritikos, 2021; Larsson, 2020), and on recommendations to follow regulations from the beginning (Floridi, 2018). The literature has discussed these aspects from a company law perspective (Hickman and Petrin, 2021). Furthermore, the importance of iteration and public discussion has been identified (Cunneen et al., 2019a), emphasizing that market forces alone cannot protect public interests (Ho et al., 2020) and calling for AI regulation (Gruson et al., 2019; Kriebitz and Lütge, 2020; McStay, 2020; Pillai and Matus, 2021; Truby, 2020; Truby and Brown, 2021; Wirtz and Müller, 2019; Zhang et al., 2021).

Additionally, Robbins (2020) has noted that responsible control and regulation of AI are not possible until there is sufficient knowledge of the nature of AI's functions, benefits, learning data, inputs and outputs, and limitations. This may justify striving for a balance between bottom-up and top-down approaches and between ethical approaches and regulations (Hermann, 2021a).

3.1.4.4 AI Programming Requirements

Considerations and approaches to programming ethics into AI are another attempt to respond to ethical and moral concerns and issues. They have emerged as a research focus in the literature. Arvan (2018) criticized previous approaches in three ways: too flexible, too inflexible, or give too much freedom to AI to understand and replicate human misbehavior. As a solution, Arvan (2018) suggested programming AI with the ability to simulate the past and future. Other scholars have also called for developing a version of imagination for AI programming (Pinka, 2021; Umbrello, 2020). Tajalli (2021) pointed out that developing AI's thinking ability would reduce the chances of AI going rogue. Because no moral theory addresses the full complexity of this issue, it has been recommended that the focus should be on avoiding immoral actions, that a pluralistic approach is preferable to a single theory (Gordon, 2021), and that various moral preferences and decisions of society should be incorporated into AI (Martinho et al., 2021b). The difficulty of resolving disagreements about ethics has led to calls (Bogosian, 2017) to develop a system that takes into account users' and stakeholders' different values in AI (McDougall, 2019; Umbrello, 2020; Umbrello et al., 2021). Moreover, Prestes et al. (2021) have suggested an approach to the first global ontological standard in this context. Such approaches require human monitoring and continuous revision, according to van de Poel (2020). In contrast to these value-based approaches, other proposals for programming ethics into AI consist of combining existing approaches (Cvik, 2021; Noothigattu et al., 2019), for example, using an extended utility function of AI to choose between multiple actions (Vamplew et al., 2018). Another approach is enabling AI to evaluate specific situations in terms of ethics (Benzmüller et al., 2020), which can also help map the flexibility of human moral judgment of a situation (Dubljević, 2020).

Fröding and Peterson (2020) suggested a design approach different from the above approaches that makes values and utility functions secondary by programming a friendly AI that mimics aspects of friendship and evolves through learning rather than through pre-programming of all aspects of an action (Fröding and Peterson, 2020). This approach has already been met with criticism because the programming of such AI would need to be constantly adapted to human actions and would require stricter, situation-specific legal regulations (Li, 2021). Additionally, designers have been criticized for making decisions about AI's ability to learn and thus incorporate their ethical views into the AI's programming (Baum, 2020; Sood, 2018). The literature also deals with artificial agents that recognize moral actions and link them to rewards (Haas, 2020), with simulation of ethical situations (Arnold and Scheutz, 2018), and with software reflecting personal users' preferences regarding ethics (Autili et al., 2019). However, a long-term approach to protect AI against manipulation has not yet been found (Osório and Pinto, 2019). Other technical and social challenges to programming requirements include establishing trust in artificial moral agents and ensuring responsible research and development and public involvement (Cervantes et al., 2020a). Thus, the need to equip machines with the ability to explain themselves has been highlighted (Khrais, 2020; Nallur, 2020). van Berkel et al. (2020) showed that users have different preferences in setting standards for programming. Awad et al. (2018) also illustrated this by studying preferences of people from over 200 countries. However, this study has been criticized for neglecting fundamental ethical principles, basic laws, human dignity, equality, and legal standards (Kochupillai et al., 2020). Stahl (2022) considered how AI innovation ecosystems can be shaped in responsible ways. Regarding such thought experiments, it has been noted that the future environment of AI and AI behavior cannot be predicted in the design phase because the design is based on incomplete information (Borenstein et al., 2019; Héder, 2020). The selected literature does not provide concrete approaches to ensure the required involvement of users (Martin, 2017) and ethicists (Gordon, 2020; Segun, 2021b), with the exception of Awad et al. (2018). However, proposals have already been made to consider human rights norms in the design of AI (Aizenberg and van den Hoven, 2020).

3.1.4.5 Human Moral Judgement and Moral Responsibility

Another research direction that emerged from the literature review was the study of human moral preferences and moral responsibility. For example, Frank et al. (2019) showed that moral preferences depend on the decision situation and the person's personal perspectives. These results are partially reflected in Wolff et al. (2019), who found that moral decisions depend on neural and psychological factors and are, for example, influenced by reaction times and subjectively perceived distress in a situation. Furthermore, the influence of moral preferences and individual characteristics on an AI's perception of a moral dilemma has been found (Rhim et al., 2021). One qualitative study demonstrated that human moral values may depend on contexts, cultures, and emotional states (van Berkel et al., 2020). Human judgments about decision situations and preferences for particular decisions by an AI depend on various factors.

In recent years, the discussion on moral responsibility for the consequences of actions enabled or performed by AI has been conducted in scientific articles. Burton et al. (2020) referred to these considerations as the responsibility gap, which arises from the risk that neither developers nor providers and users can be held morally accountable for harm caused by AI. Lauwaert (2021), in contrast, argued that this gap does not exist and that one person always bears responsibility. According to Martin (2019) and Héder (2020), this responsibility belongs to the companies and designers who developed the algorithms. Meanwhile, Rochel and Evéquoz (2021) argued that developers can only be held responsible for active misconduct and that a new definition of responsibility is required. Developers also place moral responsibility on themselves, on people providing input on the AI, on users, and on the machine itself (Orr and Davis, 2020; van der Waa et al., 2021). This was also discussed by Eiben et al. (2021), according to whom highly developed robotics further weaken human control and thus request human responsibility in this context. The literature not only discusses moral responsibility in the context of AI but also distinguishes several notions of responsibility (Coeckelbergh, 2021).

Few scholars have considered the possibility that AI might bear moral responsibility in the future (Sebastián and Rudy-Hiller, 2021; Tigard, 2021). Only strong AI meets the requirements of moral responsibility (Smith and Vickers, 2021) but only humans can bear this responsibility in full (D'Acquisto, 2020). In contrast, Hohenstein and Jung (2020) and Shank and DeSanti (2018) have shown that humans consider AI to be responsible and attribute blame to it. Thus, insight into scientists' reasoning and research findings in recent years suggests that the current literature does not provide a clear answer to questions about moral responsibility. Cunneen et al. (2019b) recognized the difficulty of providing clear answers to these questions.

3.1.4.6 Moral Status of AI

The review showed that the moral status of AI has been another area research has focused on in recent years. Gordon (2020) and Wareham (2021) and Gogoshin (2021) argued that the attribution of moral status should not be rejected in principle and that most people do not assign moral status to AI. For example, AI does not meet the criterion of sentience (Gibert and Martin, 2021), consciousness (Mosakas, 2021; Nath and Sahu, 2020), and liveliness (Gordon and Gunkel, 2022). Further, the vagueness of AI and its ethical behavior has been discussed in this context (Dobbe et al., 2021). Moreover, AI cannot uphold norms because it cannot consciously violate them in the first place (Swanepoel, 2021). The argument that AI cannot be a moral agent is also justified by the fact that it lacks emotion-driven behavior (Brożek and Janik, 2019) and ethically relevant properties (Farisco et al., 2020). Furthermore, the literature states that AI cannot be a moral agent because humans define and control AI. Therefore, AI should not be subject to rights and duties (Roff, 2019), and the attempt to develop ethical machines contradicts the concept of ethics itself (Sparrow, 2021). Arguments such as Sparrow's (2021), which rule out the moral status of AI even in the future, are countered by the possibility that a moral status could be ascribed to AI in the future (Livingston and Risse, 2019; Podschwadek, 2017). For example, the cognitive abilities of AI resemble those of other living beings that are considered moral patients (Shevlin, 2021). Additionally, hybrid human beings in the context of posthumanism are also discussed as moral agents that might be similar to biological human beings (Nath and Manna, 2021).

Although the literature shows that it is currently not technically feasible for artificial agents to replace human actions in morally relevant situations, the development of ethical mechanisms and cognitive structures for artificial agents is not seen as an optional condition but as an important and feasible one because of the interaction between humans and AI (Cervantes et al., 2020b; Mabaso, 2021; Scheutz, 2017). Further, research indicates that strong AI could experience emotions like humans and become an equal part of society in the future (Kelley and Twyman, 2020). Given the development of cognitive structures and emotions for AI, Anthis and Paez (2021) have recommended expanding the definition of those to whom humans pay moral attention to all sentient beings. With respect to the possibility of granting a moral status to AI in the future, the literature has advised caution (Formosa and Ryan, 2021; Herzog, 2021). A voiding giving AI such a status would be the most ethical decision (Bryson, 2018) because humans would have difficulty controlling and understanding the moral actions and criminal actions of such AI (Fabre et al., 2021). A moral transgression by an AI would only be effectively sanctioned if humans, rather than the AI itself, were responsible for its actions (Iphofen and Kritikos, 2021). There is also the risk of malware being used to abuse the granted moral status (Mowbray, 2021). Empirical studies have shown that people perceive AI differently depending on the situation (Hohenstein and Jung, 2020; Hong et al., 2020) and that the perception of the AI's moral status depends, among other things, on its behavior (Banks, 2020) and its visual presentation (Laakasuo et al., 2021). In contrast, other scientific studies have shown that AI is sometimes not perceived as acting morally at all (Gamez et al., 2020; Shank and DeSanti, 2018). Gupta et al. (2021a) showed that racist or gender-discriminatory decisions made by an AI were not equally questioned by all humans, and Borau et al. (2021) argued that the perceived moral status of an AI may depend on the specific design of the AI, for example, whether it takes the form of a female or male chatbot.

3.1.4.7 Importance of Law and Politics

A further research focus in the academic discourse around ethics and morality of AI was found to be the importance of the legal system in preventing or sanctioning the misuse of AI. Appropriate regulations are binding and can be legally and judicially enforced (Carillo, 2020), so legal frameworks also primarily guide AI developers (Orr and Davis, 2020). According to Kochupillai et al. (2020), AI programming in particular should be evaluated from not only an ethical perspective but also a legal one, thus ensuring compliance with legal principles and human values, dignity, and responsibilities. The high responsibility of legislation as well as the need for new legal norms, has also been highlighted in other studies (Huang, 2019; Pagallo, 2017; Schiff, 2021) and discussed, for example, with regard to the application of AI in schools (Mantovani et al., 2020). In recent years, judicial applications of AI and existing laws have been discussed and criticized (Miller, 2019). Whereas Schönberger (2019) argued in the context of AI in health care that current European laws and ethical approaches are sufficient, Miguel et al. (2020) wrote that the General Data Protection Regulation is incomplete and insufficient for the application of AI in the health care sector. According to the literature, other application areas of AI also exhibit inconsistency, a lack of scientific justification (McStay and Rosner, 2021), and a neglect of the specifics of the relevant technologies and contexts (Felzmann et al., 2019). These gaps in EU legislation, these gaps have also been identified in U.S. legislation (Truby and Brown, 2021), and the influence of large corporations on AI legislation has been criticized (Nemitz, 2018). Legislation and issues of ownership of AI-created innovations in relation to the patent system have also been discussed in the literature (Feng and Pan, 2021). E.U., U.S., and international laws do not currently provide a basis for the validity of these moral rights, and it is unclear when AI will be legally considered a personality (Miernicki and Ng, 2021). To this end, it has been argued that rights can only refer to humans (D'Acquisto, 2020), and robots are not legal persons and do not meet moral criteria (Gordon, 2020). Additionally, scholarly articles have discussed the policies of various countries (Vesnic-Alujevic et al., 2020). Criticisms include the lack of a long-term strategy (Cath et al., 2018) and the formulation of trustworthy AI (Ryan, 2020). According to Ryan (2020), this formulation does not reflect the current definition of trust, and it is not AI but organizations and individuals using AI that require trustworthiness. A scholarly examination of Chinese policy shows that it strongly promotes the development and use of AI because of its potential but disregards the negative consequences of AI's use. There is no balance between the ethical norms and standards for AI and their implementation (Roberts et al., 2021). Moreover, research discusses AI's relation to politics not only through analysis of political actions, but also through the analysis of the integration of AI in political decision-making processes (Sætra, 2020), the manipulation of elections (Kane, 2019; Landon-Murray et al., 2019), and cyber-attacks (Timmers, 2019). Possible measures are always associated with ethical challenges (Timmers, 2019). Table 3.4 provides a short overview of the main results.

Table 3.4 Key Content of the Different Research Topics

Topic	Main Content
Ethical and Moral Implications of AI	Focus of the literature is on three application areas (implications for businesses, the health sector, and the public sector), and little research has been conducted on implications of AI for private use
	Research considers implications of physical and digital applications without referring to application areas and technologies
Approaches to Ethical Handling of AI	Recommendations have to be distinguished from principles, and the literature focuses on principles (in particular in the health sector) as well as on further approaches
Demand for Governance and Regulations	Need for governance models in relation to AI development and its applications
	Need for governance of AI on a political level and need for binding regulations
AI Programming Requirements	Discussion about programming ethics into AI, about programming a friendly AI, and about rebel agents as well as further technical and social challenges in this context
Human Moral Judgement and Moral Responsibility	Different moral preferences and human judgements leading to different moral decisions
	Moral responsibility for consequences of actions enabled or performed by AI now and in the future is intensively discussed in the literature
Moral Status of AI	Moral status of AI now and in the future is controversial
Importance of Law and Politics	Legal regulations for using or programming AI in different application areas
	Ethical challenges and different policies around the world

For a better understanding of how the 262 articles were assigned to the different topics and subtopics in Table 3.3, see Table 3.5. Table 3.5 shows that the number of studies assigned to each subtopics differed significantly. Furthermore, articles were assigned to a subtopic that fit the articles best.

Table 3.5 Overview of the Studies Assigned to each Subtopic

Topic	Subtopics and Content	Paper
Ethical and Moral Implications of AI	Ethical concerns and benefits in the context of AI	(Hermann, 2021b; Luk et al., 2021; Mao and Shi-Kupfer, 2021; Pillai and Matus, 2021; Swarte et al., 2019; Truby and Brown, 2021)
	Ethical and moral implications in the context of AI in different application areas (business, health sector, public sector)	(Colombino et al., 2021; Hamilton and Davison, 2021; Hermann, 2021a, 2021b; Macnish and Fernandez Inguanzo, 2019; Pillai and Matus, 2021; Ryan, 2019)(Brady and Neri, 2020; Brown and Halpern, 2021; Fenech and Buston, 2020; Fiske et al., 2019; Goirand et al., 2021; Kempt and Nagel, 2022; Kerasidou, 2021; Kretzschmar et al., 2019; Luk et al., 2021; Martinho et al., 2021a; Montemayor et al., 2021; Mörch et al., 2021; Morley et al., 2021a; Park et al., 2019; Saheb et al., 2021; Sand et al., 2021; SFR-IA Group and CERF, 2018; Shaw et al., 2021; Stewart et al., 2021; Stokes and Amitabha Palmer, 2020) (Calvo, 2020; Chounta et al., 2021; Liu et al., 2019b; Mark, 2019; Ryan, 2019; Ryan and Anya, 2019; Sun and Ye, 2021; Swarte et al., 2019; Szocik and Abylkasymova, 2021; Vallejos et al., 2017; Wakunuma et al., 2020)
	Implications of physical products and digital products, and implications without specific application areas, uses or technologies	(Antle and Kitson, 2021; Arifin and Lennerfors, 2022; Cunneen et al., 2019a; Danaher, 2019; Ghotbi et al., 2021; King, 2020; Liu et al., 2019b; Livingston and Risse, 2019; McStay, 2020; McStay and Rosner, 2021; Milano et al., 2020; Murdock, 2018; Nigam et al., 2021; Popa et al., 2021; Reed et al., 2021; Siegel and Pappas, 2021; Smith and Miller, 2021; Szocik and Abylkasymova, 2021; Truby, 2020; Truby and Brown, 2021; Vrščaj et al., 2020; Wangmo et al., 2019; Yew, 2021)

(continued)

Table 3.5 (continued)

Topic	Subtopics and Content	Paper
	Implications of the relationship of humans to machines	(Gill, 2020; Köbis et al., 2021; Lara and Deckers, 2020; Novak, 2020; Seeber et al., 2018; Togni et al., 2021; Weber-Guskar, 2021)
Approaches to Ethical Handling of AI	Recommendations about the use of ethics in the context of AI	(Asaro, 2019; Brady and Neri, 2020; Calvo, 2020; Du and Xie, 2021; Fiske et al., 2019; Floridi et al., 2018; Hamilton and Davison, 2021; Jobson et al., 2022; Jotterand and Bosco, 2020; Kretzschmar et al., 2019; McKernan et al., 2018; Vakkuri et al., 2020; Yew, 2021)
	Principles of dealing with AI	(Abràmoff et al., 2020; Bærøe et al., 2020; Buruk et al., 2020; Chauhan and Gullapalli, 2021; Clarke, 2019; Floridi et al., 2018; Floridi et al., 2021; Gabriel, 2020; Hagendorff, 2020; Hermann, 2021a; Ibáñez and Olmeda, 2021; Jobin et al., 2019; Joerin et al., 2020; Kenny et al., 2021; Kerr et al., 2020; Larson et al., 2020; Milossi et al., 2021; Mittelstadt, 2019; Morley et al., 2021b; Peters et al., 2020; Rességuier and Rodrigues, 2020; Robbins, 2019; Ryan and Stahl, 2021; Segun, 2021a; SFR-IA Group and CERF, 2018; Stewart et al., 2021; Stix, 2021; Taddeo et al., 2021; Vakkuri et al., 2020; van Bruxvoort and van Keulen, 2021; Whittlestone et al., 2019)
	Further approaches and human rights	(Anshari et al., 2021; Bankins, 2021; Brendel et al., 2021; Burton et al., 2017; Franzke et al., 2021; Fukuda-Parr and Gibbons, 2021; Geis et al., 2019; Gerdes, 2022; Gibbons, 2021; Kelley and Atreides, 2020; Khargonekar and Sampath, 2020; Krijger, 2021; Lannon et al., 2021; Mantelero, 2018; Mantelero and Esposito, 2021; Mökander and Axente, 2021; Mökander and Floridi, 2021; Mörch et al., 2020; Salem et al., 2020; Schrader and Ghosh, 2018; Sekiguchi and Hori, 2020; Tomalin et al., 2021; Vakkuri et al., 2021; Vallès-Peris and Domènech, 2021; Vidgen et al., 2020; Wasilow and Thorpe, 2019)

(continued)

Table 3.5 (continued)

Topic	Subtopics and Content	Paper
Demand for Governance and Regulations	Models and principles to implement existing principles as a bias for ethical governance	(Janssen et al., 2020; Lepri et al., 2021; Raab, 2020; Reddy et al., 2020; Shneiderman, 2020; Winfield and Jirotka, 2018)
	Governance of AI on a political level	(Abrams et al., 2019; Almeida et al., 2021; Floridi, 2018; Liu et al., 2019b; Liu and Maas, 2021; ÓhÉigeartaigh et al., 2020; Radu, 2021; Stahl et al., 2021a; Wu et al., 2020b)
	Need for regulation of AI	(Almeida et al., 2021; Cunneen et al., 2019a; Fernandes et al., 2020; Gruson et al., 2019; Hermann, 2021a; Hickman and Petrin, 2021; Ho et al., 2020; Iphofen and Kritikos, 2021; Kriebitz and Lütge, 2020; Larsson, 2020; McStay, 2020; Pillai and Matus, 2021; Robbins, 2020; Stahl et al., 2021b; Truby, 2020; Truby and Brown, 2021; Winfield and Jirotka, 2018; Wirtz and Müller, 2019; Zhang et al., 2021)
AI Programming Requirements	Approaches to moral and moral justification as well as user and stakeholder values to be programmed into AI	(Arvan, 2018; Benzmüller et al., 2020; Bogosian, 2017; Cvik, 2021; Dubljević, 2020; Gordon, 2021; Martinho et al., 2021b; McDougall, 2019; Noothigattu et al., 2019; Pinka, 2021; Prestes et al., 2021; Umbrello, 2020; Umbrello et al., 2021; Vamplew et al., 2018; van de Poel, 2020)
	Programming of friendly AI, that can recognize moral actions	(Baum, 2020; Fröding and Peterson, 2020; Li, 2021; Sood, 2018; Tajalli, 2021)
	Integration of rebel agents	(Arnold and Scheutz, 2018; Autili et al., 2019; Coman and Aha, 2018; Haas, 2020; Osório and Pinto, 2019)

(continued)

Table 3.5 (continued)

Topic	Subtopics and Content	Paper
	Further technical and social challenges	(Aizenberg and van den Hoven, 2020; Awad et al., 2018; Borenstein et al., 2019; Cervantes et al., 2020a; Gordon, 2020; Héder, 2020; Khrais, 2020; Kochupillai et al., 2020; Martin, 2017; Nallur, 2020; Segun, 2021b; Stahl, 2022; van Berkel et al., 2020)
Human Moral Judgement and Moral Responsibility	Moral judgement of humans and human moral values	(Frank et al., 2019; Rhim et al., 2021; van Berkel et al., 2020; Wolff et al., 2019)
	Moral responsibility of developers, providers and users	(Burton et al., 2020; Coeckelbergh, 2021; Eiben et al., 2021; Héder, 2020; Lauwaert, 2021; Martin, 2019; Orr and Davis, 2020; Rochel and Evéquoz, 2021; van der Waa et al., 2021)
	Moral responsibility of AI itself	(Cunneen et al., 2019b; D'Acquisto, 2020; Hohenstein and Jung, 2020; Sebastián and Rudy-Hiller, 2021; Shank and DeSanti, 2018; Smith and Vickers, 2021; Tigard, 2021)
Moral Status of AI	Discussion about the moral status of AI	(Brożek and Janik, 2019; Dobbe et al., 2021; Farisco et al., 2020; Gibert and Martin, 2021; Gogoshin, 2021; Gordon, 2020; Gordon and Gunkel, 2022; Livingston and Risse, 2019; Mosakas, 2021; Nath and Sahu, 2020; Podschwadek, 2017; Roff, 2019; Shevlin, 2021; Sparrow, 2021; Swanepoel, 2021; Wareham, 2021)
	Moral status of AI in the future	(Anthis and Paez, 2021; Banks, 2020; Borau et al., 2021; Bryson, 2018; Cervantes et al., 2020b; Fabre et al., 2021; Formosa and Ryan, 2021; Gamez et al., 2020; Gupta et al., 2021a; Herzog, 2021; Hohenstein and Jung, 2020; Hong et al., 2020; Iphofen and Kritikos, 2021; Kelley and Twyman, 2020; Laakasuo et al., 2021; Mabaso, 2021; Mowbray, 2021; Nath and Manna, 2021; Scheutz, 2017; Shank and DeSanti, 2018)

(continued)

Table 3.5 (continued)

Topic	Subtopics and Content	Paper
Importance of Law and Politics	Importance of the legal system to enforce regulations	(Carillo, 2020; Huang, 2019; Kochupillai et al., 2020; Mantovani et al., 2020; Miller, 2019; Orr and Davis, 2020; Pagallo, 2017; Schiff, 2021)
	Legislative gaps and policies of various countries around the world	(Cath et al., 2018; D'Acquisto, 2020; Felzmann et al., 2019; Feng and Pan, 2021; Gordon, 2020; Kane, 2019; Landon-Murray et al., 2019; McStay and Rosner, 2021; Miernicki and Ng, 2021; Miguel et al., 2020; Nemitz, 2018; Roberts et al., 2021; Ryan, 2020; Sætra, 2020; Schönberger, 2019; Timmers, 2019; Truby and Brown, 2021; Vesnic-Alujevic et al., 2020)

3.1.5 Discussion and Conclusion

This systematic literature review provides a structured understanding of the current literature on ethics and morality in AI, answering research question 1. Out of 2,561 studies, 262 articles written between 2017 and 2021 were examined. The systematization of the relevant literature into research priorities also enabled the identification of research gaps and thus contributed to theoretical and practical implications, answering research question 2. The highlighted research opportunities can provide guidelines for future research to prevent repetition and bias. Table 3.6 highlights various additional future research approaches. The literature review has shown that existing technologies can be applied in a variety of ways and in different fields (Hamilton and Davison, 2021; Reiss, 2021) and can result in different implications.

3.1.5.1 Ethical and Moral Implications of AI

Considering the ethical and moral implications of AI, the literature mainly focuses on three application areas: implications for businesses (Hamilton and Davison, 2021), implications for the health sector (Brady and Neri, 2020), and implications for the public sector (Vallejos et al., 2017). Most research was conducted in the health care sector, offering a holistic overview of the application area, and offering deeper insights for discussion and future research opportunities. In contrast, only a few articles considered ethical implications of AI in education. This is an important topic in the public sector and offers several opportunities for future research (Chounta et al., 2021). Furthermore, there is a need for a detailed overview that delineates ethical and moral implications by considering AI technologies, the applications of these technologies, and the fields of application. The research focuses primarily on applications by companies or other organizations. Only a few papers have focused on the implications of AI for private use, although the rapid development of technologies in recent years (McAleenan, 2020) and the dynamic nature of ethical concepts and principles (Carillo, 2020) call for a closer examination of the implications of AI for private use. Moreover, the research focus on the user perspective should consider implications of specific applications and technologies for physical products (Yew, 2021) and digital products (Smith and Miller, 2021).

3.1.5.2 Approaches to Ethical Handling of AI

This literature review showed that there is little focus in the scientific discourse on formulating recommendations for ethical handling of AI. Mostly, they are formulated for specific application contexts (Calvo, 2020). Principles in the scientific discourse about AI mainly address AI applications in the health sector (Stewart et al., 2021). In addition, several main principles for ethical treatment of AI are discussed in the literature (Floridi et al., 2018). Researchers have discussed the different challenges of implementing ethical principles in the context of AI (Peters et al., 2020). However, these principles are not binding (Buruk et al., 2020) and academicians should focus on supporting organizations to understand existing and future guiding principles (Clarke, 2019). Moreover, several other approaches, for example, for engineers (Sekiguchi and Hori, 2020) or students (Burton et al., 2017), are introduced and discussed in the literature. Therefore, a more general and holistic approach should be derived from these different foci, such as a human rights approach. Human rights principles have already been discussed as a common aspect of the homogeneous approach to use of AI across countries (Gibbons, 2021).

3.1.5.3 Demand for Governance and Regulations

Many articles considered in the literature review called for governance or the development of governance models by scholars (Lepri et al., 2021). Further, several models in the literature have been developed to implement existing principles for ethical governance (Shneiderman, 2020). However, the non-binding nature of these principles has also been discussed, and regulations have been called for in response (Almeida et al., 2021). Thus, the importance of the role of the government in this process should be recognized and discussed in future studies (Almeida et al., 2021). Governance does not necessarily go hand in hand with regulations (Winfield and Jirotka, 2018), so increased emphasis on the need for regulation of AI is required (Almeida et al., 2021). No articles were identified that conducted an analysis of the effectiveness of existing regulations. This is an important topic for future research. Efforts by academia to provide alternative approaches and support for the implementation of principles and ethical guidelines (Ryan and Stahl, 2021) underscore the criticism of these principles and guidelines and highlight the need for their empirical testing to determine their practical applicability.

3.1.5.4 AI Programming Requirements

Another identified area of research is the requirements and approaches for programming ethics in AI. Because the different application contexts of a machine require different design approaches (Nallur, 2020), the literature is characterized by arguments by individual authors and different approaches and models. Thus, the literature points out the difficulty of formulating one moral theory to address the full complexity of AI programming requirements and highlights the need for a pluralistic approach (Gordon, 2021). Moreover, the integration of different values of people (Bogosian, 2017) in these pluralistic approaches should be mentioned in future studies. The literature lacks an overview of the applications of AI that require particular consideration of ethics and morality and of the inclusion of user preferences in discussions on programming (Awad et al., 2018). The research focus can be divided into approaches to programming (Umbrello et al., 2021) and challenges to programming (Coman and Aha, 2018). A further approach offering opportunities for future research involves friendly AIs evolving through learning (Fröding and Peterson, 2020). Researchers have criticized these approaches by raising the issue of the requirement for more situation-specific legal regulations (Li, 2021). Thus, future research should try to generalize regulations or specify regulations for different usage situations in the context of friendly AI.

3.1.5.5 Human Moral Judgement and Moral Responsibility

The subfield of human moral judgment that considers the decision-making situation of an AI is comparatively under-researched based on the number of articles. Moral preferences depend on the respective decision situations and a person's personal perspectives (Frank et al., 2019). Thus, further research on differences of personal perceptions and on how AI can be adapted to these moral preferences should be conducted. Research could also discuss the possibility of a more general framework in this context. The review of the literature on moral responsibility revealed disagreements on the distribution of responsibility (Martin, 2019; Rochel and Evéquoz, 2021) and highlighted the question of what exactly moral responsibility means. In this context the responsibility gap poses a major risk for future harm, neither developers nor providers and users are ethically and legally responsible for actions conducted by AI (Burton et al., 2020). Accordingly, future research should discuss how to deal with this gap and include legislation and governance in the discussion. Several articles examine this topic with reference to the trolley dilemma (Frank et al., 2019). However, this thought experiment does not reflect the current technological possibilities of AI and, in the sense of autonomous driving, represents a currently limited application area of AI (Cunneen et al., 2020).

3.1.5.6 Moral Status of AI

The majority of researchers currently do not assign a moral status to AI (Gibert and Martin, 2021). However, the literature partly argues that this moral status should not be rejected in principle (Gordon, 2020). Many arguments about missing consciousness (Mosakas, 2021), missing emotion-driven behavior of AI (Brożek and Janik, 2019), and the fact that AI is controlled by humans (Roff, 2019) dominate the discourse about the moral status of AI. The literature in this context is heavily influenced by arguments by individual authors and shows disagreement about whether cognitive structures should be developed for AI (Shevlin, 2021) and what arguments should be used to ascribe a moral status to AI (Gordon, 2020). In addition to the current moral status of AI, future research should focus on the moral status of strong and futuristic AI. Thus, strong AI could meet the criteria mentioned above and become morally responsible (Kelley and Twyman, 2020). This moral status should be further discussed. Principles, guidelines, and programming requirements have already been taken into account by academics and practitioners.

3.1.5.7 Importance of Law and Politics

Although approaches to implementing morality and ethics in the context of AI have been developed in the literature, only appropriate regulations are binding (Carillo, 2020), and the responsibility and impact of legislation have been highlighted in the literature (Huang, 2019). However, researchers disagree on whether current laws on morality and ethics in the context of AI are sufficient (Miguel et al., 2020; Schönberger, 2019). Future literature should seek to continue supporting legislation through discussions and development of frameworks. Thus, close cooperation between academia and legislators is necessary in this context to develop appropriate regulations in the long term. In addition, the different characteristics of AI, such as weak and strong AI, should be taken into account. In the scholarly discourse on the roles of law and policies, both legislation and policy decisions have been criticized. Strikingly, mostly E.U. or U.S.

laws have been mentioned, and the focus of research on European policy has been highlighted (Ryan, 2020; Truby and Brown, 2021). This underrepresentation of certain regions (Jobin et al., 2019) should be balanced in further research. Huang (2019) showed that enacted laws can affect human moral judgments of AI. These findings should be reviewed because they could have important implications for stakeholders in the development of AI for international markets. Table 3.6 provides a summary of future research questions.

Table 3.6 Future Research Questions

Topic	Future Research Questions
Ethical and Moral Implications of AI	Is there a difference between ethical and moral implications within AI-based programs and products for private use and corporate use? Which ethical and moral factors influence each use case?
	What are the different ethical and moral implications that need to be considered during the private use of AI-based physical and digital products (for example, smart watches, smart speakers, smartphones)?
	To what extent do ethical concerns and ethical justifications influence the use of AI by users on the one hand and companies on the other?
Approaches to Ethical Handling of AI	To what extent is there a need for complementary sector or application-specific approaches for an ethical approach to AI? Do the possibilities for implementing existing approaches differ with regard to organizational sizes and forms?
	To what extent do AI-developing companies or AI-using organizations currently take ethics guidelines and principles into account? What criteria do they use to select these principles?
	What approaches are needed to help private users deal with the ethical and moral problems of using AI?
Demand for Governance and Regulations	How should developed governance models of ethical and moral AI be evaluated for their practicality and effectiveness?
	In what areas is AI already regulated due to ethical and moral concerns? Are these regulations fulfilling their intended purposes? How do they affect the development and application of AI in the regulated areas?

(continued)

Table 3.6 (continued)

Topic	Future Research Questions
AI Programming Requirements	To what extent are some approaches to programming ethics in AI better suited to particular application contexts than others?
	Which of the existing approaches to programming ethics in AI are preferred by users? Are there differences here in terms of application contexts?
	How should the approaches to programming ethics in AI that have been developed in recent years be evaluated for their practical feasibility and effectiveness? To what extent are such approaches already being used?
Human Moral Judgement and Moral Responsibility	How do different stakeholders define the term "moral responsibility" in the discourse around AI?
	How does the perception of AI as a moral bearer of responsibility influence the demand and application of corresponding products and services?
	How do marketers adapt their communications strategies in the context of ethical and moral AI?
Moral Status of AI	What is the practical significance of recognizing that some users ascribe moral status to AI, whereas other users do not? How will the moral status of AI change in the future?
Importance of Law and Politics	What influence do regional policy debates and regional legal frameworks for ethics and morality in the context of AI have on the development and use of AI in different regions of the world?
	How do internationally operating companies deal with the legislation of different regions on the ethical development and application of AI?

Summing up, the literature review illustrated that discussions by academia about the ethics and morality of humans and AI have increased significantly in recent years. Moreover, the literature identified a lack of research in this context. The entire discourse is characterized by different research approaches and arguments, as well as by mutual criticism within the scientific community and external stakeholders. This can be attributed to the complexity of the topic. Thus, an important finding of this work is that research in recent years is not numerically tangible and cannot be easily represented in tables. Nevertheless, the systematic literature review allowed us to show how many research articles can be assigned to a research area. Articles dealing with ethical and moral implications and approaches to ethical handling of AI could be identified most often throughout the review. In particular, implications in the application areas of businesses, in the health sector, and the public sector indicated that numerous different articles on these topics exist, whereas significantly fewer articles exist on the implications of physical or digital products. Therefore, research should focus on specific application contexts as well as general approaches and frameworks in the future. Additionally, the majority of the articles covered principles and further approaches to integrating ethical aspects of AI but did not include recommendations. In the context of governance and regulations, most of the literature focused on regulations. Literature on AI programming requirements explores many heterogeneous approaches, and many papers could be identified that dealt with the moral status of AI and tried to predict how it will exist in the future. However, many articles covered several subject areas. Therefore, it needed to be considered which topic or topics were central to a paper.

Given the lack of verification of the existing results, our systematization of the literature not only provides an overview for academics but also practitioners. For example, it demonstrates overall that close collaboration between science and policy is necessary to enable governance and regulation of an ethical approach to AI. Furthermore, organizations need to understand their responsibilities in this context. Organizations should cooperate with academics to obtain a better understanding of different approaches and frameworks to implement morality and ethics in the context of AI.

Although we made several contributions to the literature, there are some limitations to be considered. Systematizing the current literature to identify research gaps and achieve a comprehensive understanding of the field can lead to a broad perspective. Because this paper focuses on the holistic approach to ethics and morality in the context of AI and investigates the current state of the literature through different databases and journals, it would be interesting to take a deeper look at individual research branches, such as the marketing sector or the health care sector. Because the aim was to gain wide knowledge in several subfields, we were not able to cover every detail. Moreover, the review concentrated on articles with a two-year impact factor of over 0.5 from selected scholarly databases. Other publication forms like books were excluded. Nevertheless, we could provide a holistic overview of the high-quality literature, creating a foundation for further research in this context.

3.2 Essay 2—Aspects Driving Customers' Intention to Use Automated Purchasing Processes

This study was published as: Selter, J. L.; Fota, A.; Wagner, K. & Schramm-Klein, H., 2023. Aspetcs driving customers' intention to use automated purchasing processes. International Journal of Retail & Distribution Management 51 (9/10), pp. 1158–1173.

3.2.1 Introduction

The Internet of Things (IoT) and the associated digitization and automation of processes are changing and penetrating more and more consumers' everyday lives (Bellis and Johar, 2020), especially as the use of AI makes it possible for IoT devices to become increasingly autonomous (Willems, 2022). For example, customers can interact with autonomous service agents, such as chatbots working with AI (Schanke et al., 2021), or customers can use health monitoring, recommendation systems and smart household products that conduct tasks autonomously (Puntoni et al., 2021). Within the last few years, smart devices have become increasingly relevant for both private households and retailing (Gao and Bai, 2014). According to the forecast, more than 41 billion IoT devices will be used in 2027 (Newman, 2020). One reason for this might be that automating processes can make everyday tasks such as shopping easier so that consumers can benefit more through greater convenience, less effort, or time savings (Bellis and Johar, 2020; Reinartz et al., 2019). On the other hand, risks can also arise when using autonomous devices and processes, such as a loss of control (Kahlert et al., 2017), the collection and use of personal data and concerns about security (Hoyer et al., 2020). Thus, different ethical and moral issues have to be evaluated by consumers when using smart devices like autonomous technologies (Macnish et al., 2019).

One example of smart devices is fully automated smart refrigerators, which can simplify the shopping process for consumers by automatically reordering groceries from retail stores (Hoyer et al., 2020). Now, smart refrigerators can have different degrees of automation (Kahlert et al., 2017) like monitoring weight and temperature (Gupta et al., 2021b), having internal cameras that can recognize groceries based on AI (Lifewire, 2021), preparing digital shopping lists for consumers based on AI or conducting repurchase orders autonomously based on previous settings that can be picked up at the store or be sent home to the consumers (Gupta et al., 2021b; Hossain and Abdelgawad, 2018). This can save consumers time and effort, as activities such as searching for products and groceries on the shelves, waiting at the checkout and driving to the retailer when buying groceries at stationary food retailers are no longer necessary, which is especially beneficial for people with physical limitations (Baier et al., 2016). Moreover, monitoring groceries with refrigerators can lead to less wastage of food by optimizing shopping lists or tracking the expiry dates of products (Grand View Research, 2019). The market size of smart refrigerators in 2018 was valued at US$ 253.9 m and is expected to grow steadily within the next years, representing a rising future market and highlighting its practical relevance (Grand View Research, 2019). Nevertheless, consumers are still slowly adopting AI-equipped technologies (Yang et al., 2017a). Although groceries and other products are increasingly purchased online (PricewaterhouseCoopers, 2021), automation of purchasing processes in a private context has not yet taken hold (Robra-Bissantz and Lattemann, 2019). Reasons for rejecting these processes could include concerns about security (Macnish et al., 2019) since consumers might feel tracked when technology accesses personal information like eating habits or cameras installed in consumers' kitchens. In addition, invisible tasks are more often rejected, being more difficult to understand by consumers (Wünderlich et al., 2015). To overcome these barriers, the customer experience is particularly important for the consumer goods industry and technology manufacturers (Gonçalves et al., 2020). While automated ordering processes are already widely established in the business-to-business sector (Henn and Lohwasser, 2020), automated purchasing processes in a business-to-consumer context are hardly considered in research (Marikyan et al., 2021). Hence, the influence and impact of smart home technologies on the customer experience are still largely unexplored (Gonçalves et al., 2020). Existing literature often addresses automated purchasing processes and IoT technologies from a technical perspective (Leung et al., 2018). Although there have been increasing efforts to analyse the consumer perspective in the context of technology acceptance (Yang et al., 2017a), there is still a gap considering consumer behavior regarding automated purchasing processes. Even less amount of research could be identified for smart refrigerators, with only a few studies focussing on the consumer (Baier et al., 2016) or technical (Gupta et al., 2021b) perspective. There is not only the question of acceptance but also of the change in the customer journey and, thus, the customer experience when using such devices. Decision stages change due to the use of autonomous purchasing processes adjusting the setting of the smart device in the pre-purchase phase, only having limited or no impact in the purchasing phase and possibly evaluating the smart device in contrast to the brand or retailer in the post-purchase phase

(Grewal and Roggeveen, 2020). In addition, Hoyer et al. (2020) note that different benefit components can play a role. Longoni and Cian (2022), for example, use experimental studies to prove that hedonic and utilitarian components have a different impact on the preference or use of automated systems. Furthermore, regarding the use of automated systems, hedonic and utilitarian consumption motives have only been researched and understood little (Li et al., 2020). For example, Li et al. (2020) considered different aspects in the context of hedonic and utilitarian purchases. Moreover, other researchers have analysed hedonic and utilitarian attributes in the context of virtual personal assistant devices (Yang and Lee, 2019).

In summary, this paper aims to investigate consumers' intention to use automated purchasing processes and the role of hedonic and utilitarian utility components of products in this context. Thus, following research questions have been formulated:

Which roles do hedonic and utilitarian utility components of products play regarding the intention to use an automated purchasing process?

How is the intention of using an automated purchasing process affected by trust, behavioral control and inherent novelty seeking?

Addressing these issues can contribute to research in different ways. While literature regarding automated purchasing processes and IoT technologies often addresses a technical perspective (Leung et al., 2018), this paper extends the literature from the consumer perspective. Furthermore, this paper increases the amount of literature of consumer research regarding smart refrigerators being highly limited due to its new character. Moreover, it extends the literature regarding hedonic and utilitarian utility components in this context and further investigates the influence of perceived trust and behavioral control for automated purchasing processes. Finally, this paper contributes to research by showing a moderating role of inherent novelty seeking in this context.

3.2.2 Literature Review and Hypotheses

Previous research on automated purchasing processes shows that technology has a strong influence on the customer journey and, thus, on the customer experience (Grewal and Roggeveen, 2020). According to the literature, customer experience can be referred to as"the internal and subjective response customers have to any direct or indirect contact with a company" (Meyer and Schwager, 2007). Furthermore, the customer journey is the interaction between consumers and retailers at various contact points or touchpoints throughout the purchasing process (Zinkann and Mahadevan, 2018). Touchpoints can be complemented by new technologies leading to the customer journey no longer being linear but jumping between different purchasing phases (Grewal and Roggeveen, 2020). For example, consumers might jump from the pre-purchase phase directly to the post-purchase phase, while the purchase phase is taken over by the technology itself (Grewal and Roggeveen, 2020). According to further literature, the pre-purchase and post-purchase phases can be penetrated by technologies such as virtual assistants and IoT that offer large amounts of information or radio-frequency identification chips that enable easy checkouts without human salespersons (Hoyer et al., 2020). By using technologies alongside the customer journey, new touchpoints can emerge, or existing touchpoints can change. IoT and AI can be used for automating transactions leading to convenience and time savings or simplifying purchasing processes, boosting the experimental value (Hoyer et al., 2020).

An example of new technologies penetrating the customer journey is the smart refrigerator. Smart refrigerators are refrigerators working with the help of IoT and AI (Gupta et al., 2021b). These devices can, for example, automatically adjust the temperature according to available groceries (Gupta et al., 2021b), recommend recipes based on consumers' preferences, track the expiring dates of groceries or identify missing and required groceries and reorder them autonomously (Baier et al., 2016). As soon as missing groceries are recognised, for example due to the weight of food in the refrigerator falling below a pre-set limit value or by cameras working on AI, the refrigerator automatically reorders the groceries and the consumer gets a delivery (Gupta et al., 2021b). Moreover, consumers can get notifications of this process via mobile applications (Gupta et al., 2021b), and the purchasing process can be accepted or rejected by consumers (Murali et al., 2017). Having the purchasing process conducted autonomously, less effort and time has to be invested by consumers in the purchasing process of groceries (Bayer et al., 2021), which could enhance their customer experience (Balaji and Roy, 2017). However, smart refrigerators are able to collect, filter, analyse and interpret large amounts of data about private information, preferences and habits of consumers in real-time (Robra-Bissantz and Lattemann, 2019). Thus, a higher degree of automation might trigger a feeling of lost control for consumers, leading to increased uncertainties and risk perceptions (Kahlert et al., 2017). This might result in barriers to acceptance of automated purchasing processes, for example, due to concerns about self-efficacy, control, privacy and data security (Wünderlich et al., 2013). In addition, customers tend to reject services that are highly invisible (Wünderlich et al., 2015). Moreover, literature has often shown that people are sceptical of algorithms, resist them and prefer other recommendations, such as those made by humans (Longoni and Cian, 2022). This can be explained by the fact that people believe that algorithms only operate according to clear rules and do not respect the individuality of humans (Bellis and Johar, 2020), which has already been tested using the example of AI (Longoni et al., 2019; Longoni and Cian, 2022). Furthermore, Table 3.7 offers an overview about the most relevant literature in the context of automated retail servicers and especially consumers' use of smart refrigerators.

Table 3.7 Overview about most Relevant Literature

Source	Focus	Topic
Baier et al. (2016)	Smart homes for automated consumer goods	Exploration of logistical integration of automated consumer goods supply and examination of consumers' attitude regarding smart refrigerators
Bellis and Johar (2020)	Autonomous shopping systems	Examination of psychological and cultural barriers to adopting autonomous shopping systems and suggestion of overcoming these barriers
Grewal and Roggeveen (2020)	Customer journey management in technology intensive environment	Point out changes of the customer journey like looping or jumping between stages due to the use of new technologies and especially automation
Gupta et al. (2021b)	Smart refrigerators	Development of model for smart refrigerators that detects quantity of food and controls temperature in combination with mobile application
Hoyer et al. (2020)	Technologies transforming customer experience	Development of framework to understand the role, new technologies have on customers' shopping journey
Kahlert et al. (2017)	IoT services in retailing	Investigation of acceptance of technological autonomy in the context of IoT retail services and identification of relevant factors for consumer acceptance

Regarding theoretical approaches, a large part of research on automated purchasing processes addresses technology acceptance using the Technology Acceptance Model (TAM) (Kahlert et al., 2017). However, the TAM cannot adequately explain the use of new technologies because it does not imply individual personality differences (Godoe and Johansen, 2012). For this reason, the theory of technology readiness, according to Parasuraman (2000), is consulted. According to Parasuraman (2000), the affinity for technology results from an interplay of positive feelings (meaning potential drivers) and negative feelings (meaning barriers to adopting and using new technologies to achieve their individual goals). These are divided into four personality dimensions: optimism, innovativeness, discomfort and insecurity (Parasuraman, 2000). Optimism refers to the personal conviction that technology and, in this case, automated processes through the smart refrigerator offer more control, flexibility and efficiency for the consumer. Innovativeness expresses the tendency to seek the adoption of new technologies and be a pioneer in the context of new technologies. Discomfort, on the other hand, describes the loss of control and oppression caused by the new technology, leading to negative feelings like uncertainties and concerns regarding new technologies. Finally, insecurity refers to distrust and scepticism about not being able to use technologies like smart refrigerators properly (Parasuraman, 2000). From these four components, a positive or negative technology affinity develops, influencing the intention to use the smart fridge (Godoe and Johansen, 2012). Influencing aspects like missing control of and trust in automated purchasing processes and seeking innovative technologies can be derived as important drivers for usage intention, according to Parasuraman (2000).

Another aspect influencing the customer experience alongside the customer journey is the distinction between hedonic and utilitarian products and consumption (Gentile et al., 2007; Sachdeva and Goel, 2015). According to the first definition by Hirschman and Holbrook (1982), hedonic consumption means that consumers experience products with multiple senses and charge them with emotional meaning. Shopping is associated with fun and joy being perceived subjectively (Botti and McGill, 2011) and can also include goods of daily use (Hirschman and Holbrook, 1982). In this case, due to the affectivity, impulse purchases often occur (Kushwaha and Shankar, 2013). Kushwaha and Shankar (2013) were able to show, based on their empirical study, that consumers prefer to use several channels when purchasing hedonic products since different channels favour the possibility of variety-seeking and impulse purchases due to a larger selection, which also contributes to the fact that consumers spend more money. Since individuals use hedonic products for identity formation (Berger and Heath,

2007; Leung et al., 2018; Reed et al., 2012; Weiss and Johar, 2013), this is more related to higher involvement in comparison to utilitarian products. Because of this higher involvement, it can be assumed that an automated shopping process, which results in lower active participation by the consumer, is more likely to be rejected in the context of hedonic consumption, as also assumed by Leung et al.'s (2018) study. In contrast, there is the utilitarian benefit component, in which the focus is on functional, efficient and goal-oriented task fulfilment (Li et al., 2020). Typical types of utilitarian products are everyday products, such as milk (Strahilevitz and Myers, 1998). According to Okada (2005), consumers evaluate the product range based on the personally perceived benefit and prefer purchasing processes that require less time and effort. Chiu et al. (2014) show that these utilitarian advantages create higher utilitarian value for consumers who express their intention of an action, for example through the further use of a smart refrigerator for an efficient purchase process. Since consumers reject automation for identity-relevant products like hedonic products (Leung et al., 2018), consumers will, instead, adopt an automated process if it contributes to a more efficient and effective satisfaction of their utilitarian needs and goals (Kuehnl et al., 2019), driving usage intention (Chiu et al., 2014). Here, the usage intention refers to the respective purchasing process. In addition, according to Hoyer et al. (2020), optimal decisions can be made using smart IoT devices, particularly for utilitarian products. Therefore, we suppose:

H1: The usage intention of the purchasing process is influenced negatively by (a) the automated shopping processes and (b) hedonic products.

Studies have shown that a decisive acceptance factor in the adaptation of smart home devices lies in the consumer's perception of security and control (Kuebel and Zarnekow, 2015). Trust is closely related to these constructs because it can help reduce uncertainty. When using automated ordering processes, consumers lose control over the purchasing process and are forced to place their trust in smart devices or the companies behind them. At the same time, an automated ordering process comes with further risks, such as data security, self-efficacy and privacy issues (Wünderlich et al., 2013) that could contribute to compromised trust (Macnish et al., 2019). Furthermore, the study by Merritt et al. (2013) shows that different variants of automated systems with regard to their performance influence the trust towards them and utilitarian and hedonic attitudes towards a product brand exert an influence on trust (Chaudhuri and Holbrook, 2002). Thus, it is expected that both the effect of the product type and the automation of the process on the usage intention are mediated by trust. Especially in the case of highly autonomous technologies, trust plays a major role in consumers' acceptance (Constantinides et al., 2017). Moreover, since the use of IoT has grown significantly, consumers' trust regarding IoT and autonomous sales processes is considered to be important for consumers' intention to use it (Constantinides et al., 2017; Wünderlich et al., 2013). Based on this, we made the following assumption:

H2: The effect of (a) the automatisation of the shopping processes (not given vs given) and (b) the product type (hedonic vs utilitarian) on the usage intention of the purchasing process is mediated by trust.

Perceived control measures how highly consumers assess their control over the respective purchasing process and their knowledge and skills for its successful execution (Gao and Bai, 2014). The Theory of Planned Behaviour (Ajzen, 1991), amongst others, has shown that perceived control is highly relevant when forming an intention because it reflects the human need for autonomy (Bellis and Johar, 2020) and self-attribution (Botti and McGill, 2011; Leung et al., 2018). When consumers perceive control, this can lead to higher satisfaction (Bellis and Johar, 2020; Botti and Iyengar, 2006; Botti and McGill, 2011), whereas decreased perceived control can evoke negative emotions (Bellis and Johar, 2020; Peterson and Stunkard, 1992). In automated purchasing processes, the IoT device takes over the main tasks of decision-making processes, reducing or diminishing control from consumers and reducing self-efficacy (Wünderlich et al., 2013). Loss of control, which is given due to the characteristics of automated (purchasing) processes, acts as an obstacle to the acceptance of such technologies. Several studies have shown that it is likely to reduce consumers' intention to use (IoT) devices (Bellis and Johar, 2020; Collier and Sherrell, 2010; Gao and Bai, 2014). Thus, it could be assumed that it is not the automated process that could lead only to a direct influence on the usage intention but the perceived behavioral control, which is influenced by the degree of automation and thus indirectly affects the usage intention. For example, Jörling et al. (2019) demonstrated that the technological autonomy of service robots limits consumers' perceived control, resulting in lower self-attribution for positive outcomes but higher self-attribution for negative outcomes. Since a smart home device like a smart refrigerator makes automated decisions, we assume that consumers estimate their control to be lower than in a non-automated purchasing process and that an increase in perceived control can leverage the usage intention:

H3: The effect of (a) the automatisation of the shopping processes (not given vs given) and (b) the product type (hedonic vs utilitarian) on the usage intention of the purchasing process is mediated by behavioral control.

When considering the adoption of new technologies and innovative services, consumers are likely to differ because of different attitudes and motivations towards using such applications (Heidenreich and Handrich, 2015a). For example, inherent novelty seeking has been identified as a main individual trait influencing the attitude towards technology (Dabholkar and Bagozzi, 2002; Heidenreich and Handrich, 2015a). Inherent novelty seeking can be understood as the seeking of new stimuli and should be distinguished from the fundamental attitude of acquiring new stimuli (Hirschman, 1980). According to Hirschman (1980), when using innovative approaches, consumers with a higher level of inherent novelty seeking have a stronger internal motivation to use these. In the context of innovative technologies, such as intelligent personal assistants, research has identified a strong influence of individual consumer differences and especially inherent novelty seeking on adoption intentions (Handrich, 2021). Other research has demonstrated a moderating impact of novelty seeking in the context of gamification within an omnichannel setting (Kim et al., 2020) or in the use of intelligent chatbots (Um et al., 2020). According to the literature, when using innovative approaches like automated purchasing processes, consumers showing a high inherent novelty seeking are more interested in using these technologies (Dabholkar and Bagozzi, 2002). Due to that, it is supposed that consumers with a high inherent novelty seeking are more likely to use automated purchasing processes, leading to the following hypothesis:

H4: The effect of the automatization of the shopping processes (not given vs given) on the usage intention of the purchasing process is positively moderated by inherent novelty seeking.

Figure 3.5 summarizes the development of the hypotheses representing the conceptual research model.

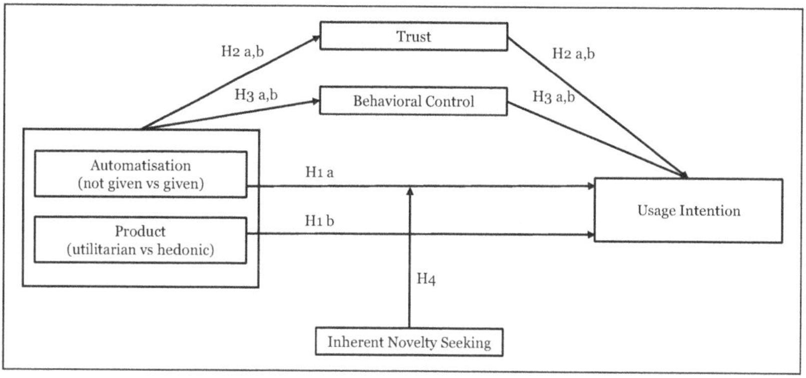

Fig. 3.5 Conceptual Research Model

3.2.3 Methodology

The hypotheses were tested via a between 2 (automation: not given vs given) × 2 (product: utilitarian vs hedonic) subject design in a quasi-experimental online study. All scenarios were depicted with a textual explanation and participants did not have to conduct the purchasing process on their own. In the "non-automated" scenarios, participants were told to open their refrigerator and identify some missing groceries. Because of that, they were told to procure these groceries in the typical way for them. In the 'automated' scenarios, participants were told they had a smart refrigerator with cameras working with AI. The application would identify missing groceries on its own and reorder them automatically to their home based on previously adjusted settings. For the manipulation of the product, a pre-test was carried out, in which the participants (N = 23) were asked to rank twelve groceries regarding the benefit component (1 = utilitarian, 5 = hedonic). Based on the mean values, milk (M = 1.91, standard deviation (SD) = .95) was selected for the utilitarian and milk cuts (M = 4.70, SD = .70) for the hedonic product. This results in four scenarios in which the participants purchase milk independently (N = 34), purchase milk automatically (N = 54), purchase milk cuts independently (N = 42), and purchase milk cuts automatically (N = 46). After performing data cleaning, which involved removing data sets with "speeder" responses and missing values, a total of 176 participants were included in the final data analysis. Of these participants, 53.4% were women and the average age was 28.63 years (SD = 11.51). The survey was distributed via social media networks and forums that deal with aspects of AI and automated purchase processes. Most participants had a university degree (44.3%) or at least a high school diploma (36.9%), whereby the minority had a monthly net income of more than €3,000 (7.3%) or less than €500 (18.8%). Many participants filled out the survey during their studies, explaining the net income of €500 to €999 (25.0%). Other participants responded that they earn between €1,000 and €1,999 (25.0%) or €2,000 and €2,999 (29.9%). Considering the purchasing frequency, the majority stated that they grocery shop once a week (42.6%) or several times a week (47.7%). The field of automated purchases is still in its infancy, so it is argued that young adults will be the next generation to face this new technology. Therefore, it is important to learn more about the potential usage intention of consumers who will potentially use this technology in the future and derive valuable insights for decision makers and future research. Established constructs from the literature were adapted for this study to test the hypotheses and queried on a seven-point Likert scale from "not apply at all" (1) to "fully apply" (7). Reliability was measured by calculating Cronbach's Alpha. Furthermore, average variance extracted (AVE) and composite scale reliability (CR) were measured, whereby the required threshold values for the three values ($\alpha \geq 0.7$; $AVE \geq 0.5$; $CR \geq 0.6$) were achieved for all constructs except behavioral control, which is only just below the limit (Hair, JR et al., 2014). The constructs used for the experimental study are presented in Table 3.8.

Table 3.8 Overview of the Constructs Used

Construct	Item	α	AVE	CR
Trust (Sheinin et al., 2011)	The purchase process shown is dependable.	.92	.78	.66
	The purchase process shown is reliable.			
	The purchase process shown is trustworthy.			
Behavioral Control (Gao and Bai, 2014)	The purchase process shown is entirely within my control.	.77	.54	.50
	I have the resources, knowledge, and ability to use the purchase process shown.			
	I can skilfully conduct the purchase process shown.			

(continued)

Table 3.8 (continued)

Construct	Item	α	AVE	CR
Inherent Novelty Seeking (Dabholkar and Bagozzi, 2002)	I am always seeking new ideas and experiences.	.81	.51	.86
	When things get boring, I like to find some new and unfamiliar experiences.			
	I prefer a routine way of life to an unpredictable one full of change.			
	I like to continually change activities.			
	I do not like meeting consumers who have new ideas.			
	I like to experience novelty and change in my daily routine.			
Usage Intention (Balaji and Roy, 2017)	Given the chance, I intend to use the purchase process shown in the future.	.95	.78	.66
	My intentions are to use the purchase process shown in the near future.			
	If I could, I would like to use the purchase process shown in the future.			

To check whether the manipulations were successful, the participants were first asked whether they had to purchase the missing food themselves in the scenario or whether it was purchased automatically from the refrigerator. All participants correctly identified the assignment (no automation vs automation). Analysis of variance (ANOVA) tests were calculated to show the successful manipulations of the utilitarian product group (Utilitarian: $M_{utilitarian} = 4.93$ (1.34), $M_{hedonic} = 2.92$ (1.42), $F(1, 174) = 94.462$, $p < .001$, $\eta 2 = .352$) and hedonic product group (Hedonic: $M_{utilitarian} = 3.49$ (1.06), $M_{hedonic} = 3.92$ (1.27), $F(1, 174) = 5.913$, $p < .05$, $\eta 2 = .033$). The constructs to collect utilitarian and hedonic product characteristics were measured according to Voss et al. (2003) (utilitarian: 5 items, $\alpha = .94$: "effective/ineffective", "helpful/unhelpful", "functional/not functional", "necessary/unnecessary" and "practical/impractical"; hedonic: 5 items, $\alpha = .88$: "not fun/fun", "dull/exciting", "not delightful/delightful", "not thrilling/thrilling" and "enjoyable/unenjoyable") by using a semantic differential.

For testing the mediating effects postulated, the SPSS macro PROCESS (model 4) was used (Zhao et al., 2010) and the four steps according to Baron and Kenny (1986) were followed. Furthermore, PROCESS (model 1) was used for testing the moderating effect (Zhao et al., 2010).

3.2.4 Results

ANOVA tests were calculated to evaluate the influence of the two experimental factors on the usage intention. For calculation, the data set was divided into two groups (0 and 1) according to the degree of automation. The results are presented in Table III and show significant differences between the "automation" vs "non-automation" scenarios. In addition, the calculation of a linear regression also illustrated this difference ($ß = -.528$, $p < .001$). For the second experimental factor, where the product differs between the two experimental groups, the data set was divided into two groups: utilitarian and hedonic product (0 and 1). However, there was no significant difference, also becoming clear when calculating a linear regression ($ß = -.034$, $p > .05$). Table 3.9 shows the results of the ANOVA.

Table 3.9 Results of ANOVA

		Automation		Product	
		not given	given	utilitarian	hedonic
Usage intention	M	4.62	2.70	3.59	3.47
	(SD)	(1.63)	(1.47)	(1.93)	(1.69)
	F-value (η^2)	67.178*** (.279)		.197 (.001)	

M: mean (SD: standard deviation); Sig.: *p < .05, **p < .01, ***p < .001

Thus, hypothesis H1a can be confirmed, as the study showed a significant negative effect of an automated buying process on customers' usage intention. However, hypothesis H1b could not be confirmed, as the expected significant impact for the second experimental factor could not be observed. Thus, consumers seem to have fewer intentions to have their food in the refrigerator bought by automated processes, while the product type had no impact.

To test the mediating effects, the direct effects of the independent variables on the dependent variable were calculated in step one, as shown in Table 3.9. Although no significant effect of the path of the product on usage intention was measured, mediation can still be calculated. Thus, the calculation of steps two and three are presented in Table 3.10.

Table 3.10 Results of Mediation Analysis

IV	DV	ß	T	F	R	R^2
Automation	Trust	−1.274***	−5.941	35.298***	.411	.169
Automation Trust	UseInt	−1.422*** .392***	−5.907 5.058	51.125***	.610	.372
Product	Trust	.133	.570	.324	.043	.002
Automation Trust	UseInt	−.199 .583***	−.834 7.531	28.489***	.498	.248
Automation	BehCon	−1.273***	−6.598	43.531***	.447	.200
Automation BehCon	UseInt	−1.429*** .387***	−5.734 4.420	46.931***	.593	.352
Product	BehCon	.019	.089	.008	.007	.000
Product BehCon	UseInt	−.133 .612***	−.552 7.168	25.820***	.480	.230

DV = dependent variable, IV = independent variable; UseInt = usage intention, BehCon = behavioral control, Sig.: *p < .05, **p < .01, ***p < .001

Considering the mediating effects and steps two and three, significant results could only be confirmed for the experimental factor automation. Thus, no mediating effects of behavioral control and trust were identified when considering the product type as the experimental factor, rejecting hypotheses H2b and H3b. Moreover, partial mediations can be shown for behavioral control on the effect of automation on the usage intention (without mediator: ß = −1.92, p < .001; with mediator: ß = −1.43, p < .001; relative indirect effect = −.49, standard error (SE) = .11, 95% confidence interval (CI)[−.73,−.28]) and for trust on the effect of automation on the usage intention (without mediator: ß = −1.92, p < .001; with mediator: ß = −1.42, p < .001; relative indirect effect = −.50, SE = .13, 95% CI[−.77,−.27]), confirming hypotheses H2a and H3a.

Testing the postulated moderating effect of inherent novelty seeking, significant results were calculated for the effect of automation on the usage intention (F(3,172) = 25.190, p < .001, R2 = .305, Int: ß = .54, p < .05), confirming H4.

3.2.5 Discussion and Conclusion

While other studies mainly focus on the technology-oriented perspective, this study further contributes to extended research of automated shopping processes from a consumer behavior perspective and extended research regarding hedonic and utilitarian components in this context. Therefore, the study shows that, despite the growing importance of digitisation and technologisation in consumers' everyday lives, new technologies, although they can make everyday life significantly easier, are not necessarily accepted and used by consumers. The findings reveal new insights that a gradual approach to a new technology, where consumers can slowly build trust and retain control over the process, can lead to its use in the long term rather than luring consumers with futuristic innovations.

Implications for Theory and Discussion

Looking at the findings in more detail, several contributions for theory can be derived. Against the assumptions that automation contributes to an efficient satisfaction of utilitarian needs (Kuehnl et al., 2019) and as a first main contribution, no differences to hedonic needs could be measured, which could also result from the fact that both products are everyday consumer goods being purchased in similar ways and represent low-priced products with a low purchasing risk. Regarding the first research question, we could not identify a role of hedonic and utilitarian utility components of products regarding the intention to use an automated purchasing process. Interestingly, the empirical study has shown that customers (still) prefer non-automated buying processes over automated ones, representing the second main contribution of this paper. One explanation could be that consumers have to engage in a new and unfamiliar purchase process and act outside their habits (Wood and Neal, 2009), outweighing benefits like convenience and time savings (Bayer et al., 2021). Furthermore, adapting automated purchasing processes can be associated with high investment costs, for example, higher acquisition costs for IoT devices (Gonçalves et al., 2020) and more initial effort, for example, when setting up the smart home device (Gupta et al., 2021b). Moreover, regarding the second research question, several other contributions can be derived from this study. Perceived trust and behavioral control have been identified as positive mediators in the automatisation of the buying process. Both factors are influenced negatively by automated buying processes. Results are consistent with previous literature that identify a lack of trust (Macnish et al., 2019) in automation processes due to perceived data security risks and uncertainties about privacy (Wünderlich et al., 2013). Furthermore, results regarding behavioral control are in line with Gao and Bai (2014), pointing out behavioral control as the main driver for users' intentions in the context of IoT. Previous research has shown that decreased control evokes negative emotions (Bellis and Johar, 2020) and the technological autonomy of robots limits customers' perceived control, leading to negative expectations (Jörling et al., 2019).

Building on the findings, theoretical implications can be concluded: the Uncanny Valley effect describes how the acceptance of robots increases up to a certain degree of anthropomorphism before dipping steeply because people get uncomfortable when robots become too human-like (Boradbent, 2017). Accordingly, an automated purchasing process could be perceived as positive until the process becomes too autonomous and artificially intelligent and acceptance drops precipitously. In this experimental study, the purchasing process could have represented such a high degree of automation that consumers already perceived this negatively, according to Uncanny Valley (Mori et al., 2012). This leads to the consideration of whether the Uncanny Valley can take place not only in a visual-human-like robot but with a process executed by AI.

Implications for Practice

Furthermore, implications for practice can be derived: When they intend to support automated shopping processes, retailers should adopt strategies to keep switching costs as low as possible and create an optimised customer experience. The customer experience has been enhanced through a novel approach to grocery shopping that facilitates a more convenient, rapid, and autonomous process, thereby improving overall quality of life. With the fast-paced and dynamic nature of daily life for many consumers, transparency about food ingredients is becoming increasingly challenging and grocery shopping can be a significant hurdle. However, the integration of the IoT into consumers' homes has the potential to not only simplify life but also encourage the purchase of healthier products through active nudging. Also, companies should strengthen consumers' perceived trust and control, for example, through consistent consideration of transparent data protection aspects (Javaria et al., 2020), playing an important role in the use of smart home devices (Yang et al., 2017a). Also, in the case of the type of device that was studied, i.e. a smart refrigerator, it is necessary that the food is delivered satisfactorily, is kept cool for a long time and is of a consistently appropriate quality, which consumers can no longer check themselves, such as in a brick-and-mortar store (Baier et al., 2016). Related to increased trust and a higher perception of control is the implementation of ways in which consumers can influence the buying process so that their scope of action is expanded in connection with automation. In the case of the device studied, this implementation, for example, could be realised by the option to add groceries to the shopping list via a touchpad on the refrigerator or an app. Alternatively, an option to make a manual change to the brand preference could be added, which would take into account variety-seeking motives and eliminate uncertainties regarding the lack of control possibilities of technology-driven automation (Lim, 2003). Additionally, this study could show the moderating effect of novelty seeking in the context of automated shopping processes. Thus, companies offering these approaches and corresponding technologies should distinguish between consumer types with different attitudes and motivations (Heidenreich and Handrich, 2015a). When focusing on an innovative concept, reaching out to customers that seek novelties and are open-minded and internally motivated might contribute to success in the starting phase (Hirschman, 1980).

Limitations and Further Research Agenda

However, it must be noted that this work also has some limitations. Research has addressed that when automating shopping processes, consumer experiences change along the phases of the customer journey (Wünderlich et al., 2013). This study, however, did not analyse the customer journey but only focused on general aspects influencing usage intentions. However, automation reduces the need for consumers to be involved in the purchasing process, eliminating the need for them to actively engage in the selection, payment and procurement of groceries and representing a change in the customer journey (Hoyer et al., 2020). Regarding this empirical study, the purchase process in the experiment was presented through full automation, while different degrees were not considered. Additionally, the empirical study focused on reordering groceries based on adjusted settings. However, associated processes like payment or delivery were not considered.

The limitations of this study show that there is potential for further research. We suggest five main research directions that tie in with the present study: 1) Different degrees of automation (Kahlert et al., 2017) should be considered, checking for an Uncanny Valley effect in the context of automated buying processes (Mori et al., 2012). 2) Future research should distinguish between reordering and purchasing processes and investigate related aspects such as payment and delivery from both consumers' and companies' perspectives. 3) Furthermore, the automated purchasing process is also logistically relevant (Robra-Bissantz and Lattemann, 2019), thus, aspects regarding the supply chain should be included and investigated in future studies. 4) Ethical issues such as the access of smart devices to payment information (Gupta et al., 2021b) or further potential privacy issues related to profiling and manipulation of automated decision-making should also be considered (Bayer et al., 2021). 5) Finally, exploring how automated purchasing processes affect consumer relationships and customer loyalty with food retailers warrants further investigation.

3.3 Essay 3—Transferring Trust, Risk and Security on Users' Intention to Trade Cryptocurrency, Using PayPal as an Example

This study was published as: Fota, A.; Selter, J. L. & Schramm-Klein, H., 2022. Transferring Trust, Risk and Security on Users' Intention to Trade Cryptocurrency, Using PayPal as an Example. Proceedings of the 2022 ECIS, 119.

3.3.1 Introduction

Online commerce is booming—and with it digital payments and transactions made online. With 361 million users, PayPal is considered to be one of the largest online payment service providers worldwide (PayPal, 2020b). When PayPal announced that cryptocurrencies could be bought and paid via the platform within the USA, Bitcoin price rose to an annual high immediately. Since 2021, PayPal has also offered its customers in the USA the option of using cryptocurrency as a means of payment for payment transactions, while the seller receives dollars (Sparkes, 2021). PayPal is also planning to enter the cryptocurrency market in other countries in the coming years (Handelsblatt, 2020), having started in Europe in August 2021 by offering customers in the U.K. to buy and sell cryptocurrencies (Browne, 2021). Via simple explanations in video form as well as seamless integration into the in-house app, which is already used daily by millions of customers around the world, PayPal is trying to make cryptocurrencies accessible to more and more customers (PayPal, 2020a). One month after the launch in the USA, already one-fifth of PayPal's U.S. users had used the PayPal app to buy or sell Bitcoins (Reynolds and Voell, 2020), which is one of the four via PayPal tradable cryptocurrencies along with Ethereum, Bitcoin Cash and Litecoin (PayPal, 2020a). Users who have already purchased cryptocurrencies through PayPal have logged into PayPal twice as often as before the purchase, generating 12% more revenue (PayPal, 2021a). For PayPal, introducing the ability to use cryptocurrencies through their platform seems to have paid off.

The trend towards the use of cryptocurrency and especially the possibility to use cryptocurrency is not only shown by the example of PayPal. Digital currencies are increasingly being positioned in the public domain, having payment card providers Visa and Mastercard supporting selected cryptocurrencies on their own networks in the near future (Dhamodharan, 2021; Visa Inc., 2021), individual companies, such as Tesla or Microsoft, already accepting payments with cryptocurrency in some cases (Sparkes, 2021), and the European Central Bank already working on a digital e-euro (Jackisch, 2021). However, the acceptance of cryptocurrencies is dwindling and German consumers, for example, increasingly know more about cryptocurrencies, but use them less (Bock, 2020).

When analyzing cryptocurrency acceptance, trust seems to play a crucial role (Arli et al., 2021). We therefore refer to the concept of trust transfer as it is described by Stewart's (2003) Trust-Transfer Model as the basic theoretical framework of this study. Previous studies have shown that the rejection of the use of cryptocurrency results, among others, from a low level of trust in cryptocurrency itself (Arli et al., 2021) as well as from a low level of trust in the blockchain technology (Völter et al., 2023). In this context, trust can be described as institution-based trust. Institution-based trust plays a central role in building trust in a business environment with which there is no prior interaction (Zucker, 1986). Hereby, institution-based trust is linked to the existence of third-party structures and trust in an institution like PayPal also increases users' trust in sellers distributing via this marketplace (Pavlou and Gefen, 2004). According to previous research, trust in cryptocurrencies is influenced by various aspects, such as being traded in a decentralized location (Arli et al., 2021), or transactions' transparency (Sas and Khairuddin, 2017). Moreover, trust in online services increases, when users believe that the online service has security measures built into the website (Gefen et al., 2003). Further, literature shows that attributes such as trust, which are attributed to an online service provider, can also be transferred to cooperating providers, companies, or services (Cao et al., 2018; Stewart, 2003; Wang et al., 2013). For instance, studies have confirmed that users transfer trust from a website they know to an unknown website (Stewart, 2003). Furthermore, unknown websites can gain users' trust if they cooperate with well-known brands (Delgado-Ballester and Hernández-Espallardo, 2008). In addition to trust, perceived risk has also been shown to be a critical factor in cryptocurrency acceptance and use (Mendoza-Tello et al., 2019). Most consumers seek to avoid risk (Mendoza-Tello et al., 2019), which plays a particularly crucial role in the adoption of new technologies (May So et al., 2005). In the context of cryptocurrency, risks that can be identified are volatility (Mendoza-Tello et al., 2019) or the risk of unsafe transactions due to the risk of dishonest traders (Sas and Khairuddin, 2017). As a further determinant, security concerns also play a major role in the context of cryptocurrencies (Rehman et al., 2020). In this context, financial intermediaries guarantee both security for the customer's account and financial transactions (Marella et al., 2020). Concerns about the security of cryptocurrency for example may result from concerns about cyber-attacks (Marella et al., 2020), whereby these attacks can be carried out as wallet attacks on customers, as DDoS attacks on networks, or as mining attacks (Conti et al., 2018). In contrast, security concerns can be reduced, for example, by providing information about the provider's technology infrastructure (Rehman et al., 2020). Therefore, in addition to trust, risk and security are also crucial determinants for the use of cryptocurrency.

Furthermore, literature has shown that trust in an online marketplace provider such as Amazon can also increase users' trust in sellers distributing via this marketplace (Pavlou and Gefen, 2004). Since, as described above, the determinants of risk and security are also linked to the institution, they follow the same process of establishment of institution-based trust. Moreover, previous research has shown the transfer of other determinants than trust following the Trust-Transfer Model (Teubner et al., 2019). Accordingly, a transfer of the attributes risk and security is also very likely according to the Trust-Transfer Model. Based on this, with our study we analyze whether a transfer of attributes that are associated with a payment service provider to cryptocurrency is possible and whether attribute transfer influences user intention to use cryptocurrencies. To gain insight into this topic, we use PayPal as a case study, because, for example, Kalinic et al. (2019) have shown in a review study that perceived trust, risk or security are among the main drivers of customer intention to use mobile payment systems, like PayPal. PayPal is considered the most secure payment method on the internet by 70% of users, according to a study by Seidenschwarz et al. (2020). A representative study by Cofinpro AG (2018)found that 54% of respondents would trust PayPal as a platform provider for bundling all accounts in one place (Cofinpro AG, 2018). Thus, PayPal is of significant importance as a brand, and we regard it as a suitable example for a payment system. We address the following research questions:

1) Is it possible to transfer perceived trust, risk and security in / of PayPal to cryptocurrencies when handled by PayPal?
2) Do these transferred attributes have an influence on the intention to a) use cryptocurrencies if handled by PayPal and b) use cryptocurrencies in general?

Thus, this paper investigates whether a payment method that is predominantly perceived as risky by many users is potentially more likely to be used by consumers if the payment service provider offering this payment method is already trusted by consumers. To the best of our knowledge, an extension of a payment service provider's offering such as trading cryptocurrencies via PayPal has not been studied before.

The results of this study are expected to add value to further research in IC in several ways: By answering the research questions, implications for theory can be derived from this study. We apply the Trust-Transfer Model in a context, which is becoming increasingly important for IC research, i.e. online payment services. In contrast to previous research, we focus on trust transfer from a service provider (company) to a payment option (service). Besides applying the model in an expanded context, we further investigate, whether an extension of the Trust-Transfer Model is possible, expanding the Trust-Transfer Model with further determinants (risk and security). These two extensions of the model provide new theoretical insights and broaden the understanding of IC-service usage in an era where technologies, business, and services are increasingly connected to each other, and new interfaces and intersections are created.

Moreover, this study also creates added value for practical applications. Different cryptocurrency providers, also including social media platform companies (for example, Meta) or state-owned central banks (for example, China's) can better understand the transfer of perceived attributes when working with payment service providers, which for example, also include traditional banks, as well as enhance their knowledge of how users' intention to use cryptocurrencies can be influenced. In addition, user intention to use cryptocurrencies and thus the cryptocurrency market itself can be strengthened. By offering more cryptocurrencies via service providers or players that are optimally already established and highly trusted as well as perceived to have low risks, more users might dare to trade cryptocurrencies, giving them access to additional payment and investment opportunities.

In the following, we introduce our conceptual framework and present our hypotheses on the transfer of trust-related attributes between a service provider and the services it trades and the influence on user intentions. By using PayPal as an example, we conducted an online questionnaire that queries the determinants of "trust", "security", and "risk" in relation to the use of PayPal in general, as well as in relation to trading cryptocurrencies via PayPal. In addition, the general usage intention of users to trade cryptocurrencies as well as to trade cryptocurrencies using PayPal is examined using a regression analysis. Our results show how users behave when a financial service provider they have known and which they trust (here: PayPal) enters cryptocurrency trading and that trust and related attributes they associate to this financial service provider are then also transferred to a cryptocurrency. Finally, we discuss our results and provide a conclusion.

3.3.2 Conceptual Framework

In the 1990s, a new method for peer-to-peer (P2P) payment transactions was introduced, using cryptography and information technology and not requiring a bank authorization process (Clemons et al., 1996). Cryptocurrencies were finally materialized with Bitcoin, a cryptocurrency still attracting the most attention within this context being introduced by Nakamoto (2008). Bitcoin is an electronic financial mechanism which is based on a decentralized organizational structure. The creation of "money" in the Bitcoin system is realized through an open algorithm, which facilitates the reliability of expectations regarding future money supply. This technology enables the real-time monitoring of transactions via a public peer-to-peer network. The entire transaction history is stored in a "chain" of transactions, often referred to as a blockchain. The blockchain represents all verified and valid transactions between users of the network. Bitcoin is therefore independent of a bank authorization process (Nakamoto, 2008). Researchers argue that cryptocurrencies rearrange the existing structure of payment and currency systems, as they are constantly gaining users' attention (Glaser et al., 2014).

Prior research on cryptocurrencies initially focused on the decentralized infrastructure and its risks. Various techniques based on network theory were used to analyze transaction anomalies and possibilities for de-anonymizing individual entities in the blockchain (Reid and Harrigan, 2013). Furthermore, Karame et al. (2012), as well as Clark and Essex (2012), focused on anonymity and transaction speed. Through a quasi-simulative approach, the authors were able to identify various deficiencies, which were then taken into account in further cryptocurrency generations. Brezo and Bringas (2012) emphasize that Bitcoin is particularly susceptible to speculation and misinformation. In addition, they criticized the lack of regulatory supervision of the market. They argue that, in general, currency exchanges are an integral part of the cryptocurrency ecosystem as they link traditional currencies to cryptocurrencies. However, as they are not regulated, they have substantial default risk.

Usually, the value of a currency is based on macroeconomic conditions, inflation and interest rates (Meese and Rogoff, 1983). However, cryptocurrencies do not offer the same features as traditional currencies, where interest rates are provided by central banks. Therefore, valuation models based on a particular interest rate are difficult to adapt. Users tend to determine the value of a cryptocurrency themselves by analyzing news or web resources. Glaser et al. (2014) conclude that most users have so far traded cryptocurrencies like assets on a stock market, rather than using them as a payment instrument. However, cryptocurrencies can now also be increasingly used as a means of payment, such as at Microsoft or Tesla (Sparkes, 2021). Furthermore, Kristoufek (2015) has shown that there is a strong correlation between users' information gathering and the price of a particular cryptocurrency. Furthermore, trust plays a crucial role for payment systems and financial transactions, with all parties involved expecting a secure transaction. For novel technologies as payment options, such as cryptocurrencies, trust is therefore particularly important (Nelms et al., 2018). In general, trust has been identified as a key driver, when trying to establish new technologies (Kumar et al., 2021; Ofori et al., 2017). Researchers investigating users' trust in cryptocurrencies have been for example Arli et al. (2021). In their study, they showed that various factors, such as knowledge about cryptocurrency, regulation by the government, or the speed of transactions can influence the trust of users in cryptocurrencies (Arli et al., 2021). Moreover, several other studies investigated the user perspective of cryptocurrency, also highlighting the importance of trust in this context. For example, Völter et al. (2023) have shown that end-user trust in blockchain technology is a crucial factor for the value proposition. The authors also emphasize the need for further research in this context. Another study by Marella et al. (2020) analyzes trust in cryptocurrency using Bitcoin as an example and identifies eleven attributes that have a significant impact on trust in the digital currency. In addition to factors such as openness and immutability, these also include security (Marella et al., 2020). Sas and Khairuddin (2015) also look at users' trust in Bitcoin and various characteristics such as decentralization or irreversible transactions as factors influencing trust in cryptocurrencies. In addition, the researchers discuss the role of risk in dealing with cryptocurrency, such as in the context of dishonest traders (Sas and Khairuddin, 2017). Another study focused on how trustworthiness of blockchain platforms can be communicated to end users and examined various design elements (Zavolokina et al., 2019).

In addition to research about the users' perspective on cryptocurrencies, research has addressed the acceptance and usage intention of online platforms and specifically online payment systems in recent years. While Pavlou (2003) identified trust and perceived risk as important predictors of user acceptance of e-commerce, these factors have also been identified as important aspects of online payment usage intention (Nguyen and Huynh, 2018; Yang et al., 2015). Online payment is a process in which the medium of the Internet is used to make payments or manage finances and can be interchangeably used with e-payments (United Nations Conference on Trade and Development secretariat, 2001). One of the most established platforms for online payments is PayPal that was introduced more than 20 years ago. Private individuals as well as companies can transfer money via PayPal without sharing further financial information except of the email address. PayPal is therefore used as a payment option for many online stores (PayPal, 2021b). In addition, PayPal can also be used as a mobile payment app (Liu et al., 2019c), which even accounts for more than half of PayPal's usage volume. Mobile payment is understood as the payment of goods, services or bills through mobile devices such as smartphones (Dahlberg et al., 2008). Liu et al. (2019c) conducted a meta-analysis including 61 studies looking at different factors that influence the usage patterns of mobile payment methods. The authors found that, particularly in countries with Western cultures, aspects such as perceived risk and trust have a major influence on the intention to use mobile payment options. In addition, a significant positive influence of perceived security on the intention to use mobile payment was demonstrated Liu et al. (2019c). A meta-analysis by Kalinic et al. (2019) also identified the mentioned influencing factors as important explanatory variables for the intention to use mobile payment systems.

3.3.2.1 Trust

According to Chen (2006), the definition of trust in literature can be distinguished according to two schools of research. On the one hand, trust is defined as a belief, confidence, attitude, or expectation about the trustworthiness of another party. The other side holds that trust is "a behavioral intention or behavior that reflects a reliance on a partner and involves vulnerability and uncertainty on the part of the person who trusts" (Chen, 2006, p. 200). Furthermore, when considering trust, the relationship between trust and distrust and corresponding unidimensional and bidimensional models should be mentioned (Moody et al., 2017). According to the unidimensional approach, trust and distrust can be seen as one component, each representing one end of a continuum (McKnight et al., 2002). In contrast, the bidimensional approach defines trust and distrust as different constructs with different valences (McKnight and Choudhury, 2006). A hybrid approach is provided by Moody et al. (2017), who discuss that trust and distrust can coexist as independent components and that their continua are connected. In the context of our study, only trust is considered, being treated as an independent construct. Benevolence from the seller was identified as a main factor in generating trust (Moody et al., 2017) and trusting beliefs are further based on the attributes of integrity and competence (McKnight et al., 2017). Trust plays a crucial role in the context of information and communication technology, as the information levels of the various actors are often imbalanced. In online retailing, the temporal and spatial distance between the buyer and seller, which is unavoidable especially in cryptocurrency trading, increases information asymmetries and the associated risks, as well as the value of trust (Koh et al., 2012). When trading cryptocurrency, users must trust not only the online platform, but also the technology behind it. Ofori et al. (2017) published a study considering trust in an internet banking context and showing that significant predictors of trust can be information quality, service quality, privacy or security concerns. Also referring to trust transfer, Gong et al. (2020), looking at one of the largest Chinese third-party online payment platforms, have shown that trust in web payment can be transferred to trust for mobile payment (for both cognitive trust and emotional trust).

In addition to the transfer of trust, the influence of trust on intention formation also plays a significant role. Hence, trust in a trader or an online platform has been shown in previous studies to be an important indicator of users' intention to purchase (Boeuf and Sénécal, 2013; Kim et al., 2011). A study by Pavlou and Gefen (2004) confirmed that trust in the provider or an institution, such as PayPal, also increases users' trust in sellers distributing via this marketplace. This institution-based trust is the sociological dimension of trust and is concerned with structures that make the environment appear trustworthy (Keen et al., 1999). The perception of these structures of an institutional environment can influence behavior and attitudes toward a seller (Keen et al., 1999). Pavlou and Gefen (2004, p. 37) define institutional-based trust as "a buyer's perception that effective third-party institutional mechanisms are in place to facilitate transaction success." Former research identified trust to be a key driver for the user acceptance and adoption of blockchain technology and thus cryptocurrency (Fleischmann and Ivens, 2019). Moreover, trust in cryptocurrencies is enhanced, when traded in a centralized location (Arli et al., 2021).

3.3.2.2 Trust-Transfer Model

We refer to the Trust-Transfer Model introduced by Stewart (2003). Since many companies on the internet have problems generating trust among potential users, Stewart has developed a model to address this problem referring to a cognitive process. The Trust-Transfer Model describes how trust is transferred on the internet from a company already known to the user to a company or part of a company yet unknown to him. The number of links to a previously unknown website by a website known to the user increases the perceived interaction and perceived similarity between the two sites. A higher number of links between the websites leads to a greater perceived similarity by the users (Stewart, 2003). According to Stewart (2006), this link between the website of a well-known and a less well-known company can lead to an increase in trust on the website of the less well-known company. At the same time, the model also warns of negative consequences related to trust on the website of the more well-known company.

In addition, signals that lead the user to conclude that an unknown and a known website interact with each other increase trust in the unknown website and the intention to use it (Stewart, 2003).

While Stewart considers the transfer of trust on the internet from company to company (Stewart, 2003, 2006), further research has already shown that the model can also be applied to other contexts. For example, Belanche et al. (2014) confirmed that in the context of public administrations, there is a transfer of trust from provider as well as technology characteristics to electronic services, such as voting or license renewal. In addition, research by Wang et al. (2013) has shown that trust in web services can be transferred to trust in new, mobile services. Further, the transfer has been confirmed by other research and has already been demonstrated for financial service providers. While a study by Lee et al. (2007) was able to measure a transfer of customer trust in offline banks on its online banking and thus in the context of multiple channels, two further studies have also demonstrated the transfer in the context of cross-channel services. Both Lu et al. (2011) and Cao et al. (2018) conducted studies, confirming a transfer of trust from online payment to mobile payment, concluding that the transfer of trust increases the intention to use mobile payments. Since literature shows that the Trust-Transfer Model can be applied to financial service providers, in more detail online payments and mobile services, as well as in different contexts, we also adapt this model to the context of our study. PayPal's new service of now being able to trade cryptocurrency describes an extension of PayPal services that users are already familiar with. Thus, we use the Trust-Transfer Model to represent the transfer of trust from PayPal to PayPal's new service of trading cryptocurrency. In addition, we examine whether this model can also be extended by two trust-related attributes. i.e., perceived risk and security.

In the context of this study, according to the Trust-Transfer Model, we therefore argue that trust in the institution PayPal (trust in source) also increases trust in trading cryptocurrencies via PayPal (trust in target), which in turn positively influences the intention to use cryptocurrencies via PayPal, as well as in general. In doing so, we focus on the trust in PayPal with regard to its previous function as an online payment service provider for the payments of medium and small amounts via national currencies. In contrast, cryptocurrency trading via PayPal represents a service that was not possible before and is therefore new to customers. Thus, the following hypotheses arise:

H1: Perceived trust in PayPal has a positive effect on perceived trust in cryptocurrencies when handled by PayPal.

H2: Perceived trust in cryptocurrencies when handled by PayPal has a positive effect on the intention to use a) cryptocurrencies when handled by PayPal and b) cryptocurrencies in general.

3.3.2.3 Perceived Risk

Referring to the Trust-Transfer Model, previous research assumes that other attributes that are related to trust, however, are not congruent to trust, should also be transferable between different objects (Teubner et al., 2019). In addition to trust, perceived risk has been identified to have a significant influence on the intention to use cryptocurrencies (Mendoza-Tello et al., 2019). Perceived risk is associated with uncertainties of behavioral or environmental nature (Pavlou, 2003) and is used in many studies to explore users' purchase and trading behavior. Risk is defined as "consumer's perception of uncertainty" and is therefore the uncertainty about possible disadvantages of a decision (Bhatnagar and Ghose, 2004, p. 1353). Uncertainty can be understood as the potential negative consequences of buying a product or service, or for example, as in this case, as unforeseen price developments in cryptocurrency trading. Nevertheless, users, especially when trading cryptocurrency, are aware of risks such as hacker attacks and, therefore, they may feel insecure about the transmission of personal or financial data (Hong and Cha, 2013). In addition, the volatility of cryptocurrencies is seen as a potential risk and a reason why cryptocurrencies are not used (Cheah and Fry, 2015). Volatility of a currency can be described as the inability to keep the prices or the rate of the currency stable (Ciaian et al., 2016). Also, many cryptocurrencies have so far tended to be used via unknown websites and providers, resulting in a higher perceived risk. Also, perceived risk of online payments is an important aspect influencing users' behavior of online payment (Yang et al., 2015). In this context, we assume that perceived risk that users attribute to the known provider PayPal is transferred to the new service, here trading cryptocurrencies. Consequently, if users attribute a low risk to the use of PayPal in general, it can be assumed that these same users also attribute a low risk to cryptocurrencies when they are used via PayPal.

Furthermore, perceived risk exerts a negative influence on the intention to use a website (McKnight et al., 2002). Prior research in the context of cryptocurrency has shown that perceived risk negatively influences the behavioral intention (Gil-Cordero et al., 2020). Mendoza-Tello et al. (2019) identified a significant negative influence of perceived risk on the intention to use cryptocurrency, too. For example, cryptocurrency is seen as fraudulent and speculative (Arias-Oliva et al., 2019) due to its volatility (Mendoza-Tello et al., 2019). Thus, perceived risk in using cryptocurrencies via PayPal is believed to exert a negative influence on the intention to use cryptocurrencies via PayPal. Furthermore, it is believed that the perceived risk of cryptocurrencies when using PayPal also negatively influences the overall usage intention of cryptocurrencies in general. We thus suppose:

H3: Perceived risk using PayPal has a positive effect on perceived risk of using cryptocurrencies when handled by PayPal.

H4: Perceived risk using cryptocurrencies when handled by PayPal has a negative effect on the intention to use a) cryptocurrencies when handled by PayPal and b) cryptocurrencies in general.

3.3.2.4 Perceived Security

While, on the one hand, risks associated with information and communications technology are important, they cannot be considered independently of security aspects. Therefore, perceived security corresponds to the subjectively perceived probability with which a user assumes that his or her personal data and information cannot be read, recorded, or manipulated by unauthorized third parties during transmission over the Internet and storage on servers (Flavián and Guinaliu, 2006). In online payments, personal data must be disclosed in order to complete the payment process. However, when trading cryptocurrencies, payment transactions are usually encrypted and appropriate security measures are taken into account. Nevertheless, many users perceive trading cryptocurrencies as insecure (Rehman et al., 2020). On the other hand, problems with security are perceived by users in terms of product or service risk. This may arise from a lack of information about the product or service, as well as poor quality and different expectations. In this context, Rehman et al. (2020) have shown that information about the technology information of the provider of cryptocurrency can provide security to users. Thus, the main perceived risks related to cryptocurrency trading are security risks, such as also third-party service failure, user error (for example, related to illegal activities), privacy loss, and counterparty fraud. Furthermore, security attacks could be carried out as wallet attacks on users, DDoS attacks on networks, or mining attacks (Conti et al., 2018). Chellappa and Pavlou (2002) show that perceived security depends on the perceived level of encryption, the perceived protection of data, and the perceived hurdle of authentication. Referring to the Trust-Transfer Model, it can be assumed that the perceived security that the user attributes to the known website is transferred to the unknown website, too (Stewart, 2003). The transfer of the perceived security in this case is from an institution, more precisely from PayPal to cryptocurrency. In this case, the transfer will not happen automatically, but will follow the same processes of establishment of institution-based trust as has been the case with online marketplaces (Pavlou and Gefen, 2004).

In addition, security aspects contribute to better functionality of (payment) services (Tandon et al., 2018). Tandon et al. (2016) also confirmed users' security and privacy assessments as crucial parts of their opinion and behavior formation. The perceived security positively influences the intention to use a website (McKnight et al., 2002). Consequently, in the considered case, it can be assumed that the perceived security of cryptocurrencies when using PayPal positively influences the intention to use cryptocurrencies via PayPal. Furthermore, it is hypothesized that the perceived security of cryptocurrencies when using PayPal positively influences the intention to use cryptocurrencies in general. Considering earlier research work as well as the Trust-Transfer Model, we thus summarize:

H5: Perceived security of PayPal has a positive effect on perceived security of cryptocurrencies when handled by PayPal.

H6: Perceived security of cryptocurrencies when handled by PayPal has a positive effect on the intention to use a) cryptocurrencies when handled by PayPal and b) cryptocurrencies in general.

Figure 3.6 summarizes the presented hypotheses H1-H6b.

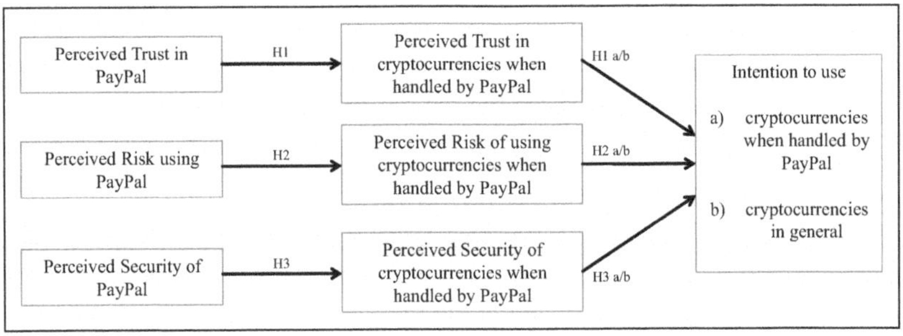

Fig. 3.6 Research Model

3.3.3 Methodology

To test the hypotheses of this study, we conducted an online study. We developed an online questionnaire, which was distributed in two different ways. On the one hand, the questionnaire was distributed on an online platform and via social media. On the other hand, the questionnaire was distributed via a larger social media channel that also frequently deals with financial topics. Thus, it can be assumed that the questionnaire was primarily distributed to subjects interested in finance. To ensure a suitable sample and control for possible influences, subjects were asked what prior experience they had in dealing with cryptocurrency. Also, the questionnaire included a general explanation of what cryptocurrencies are as well as their application at the beginning of the study, to ensure that all participants have the same understanding and knowledge of cryptocurrencies. Since Bitcoin is the best-known cryptocurrency, it was also listed as an example currency, so that the implications of this study can generally be related to this currency. Afterwards, the participants were shown an explanatory video which briefly explains the concrete use of cryptocurrency via PayPal for illustrative purpose. For this purpose, a video published by PayPal itself for the American market (PayPal, 2020a) was supplemented with German subtitles for the survey in Germany.

Prior to further analysis of the data, data cleaning was performed to eliminate incomplete or erroneous and implausible data. In addition, "speeder" participants were eliminated, i.e., participants who needed less time than half of the median to answer the questions, so that it could be assumed that the participants concerned had not dealt with the questions intensively enough (Schendera, 2011). After cleaning, the data set comprises N = 534 random participants. A large proportion of participants (87%) have already used PayPal, while cryptocurrency was used by fewer participants (14%). Of the cryptocurrency users, most have already used PayPal (90%). However, we only included participants in our study, who have already used PayPal, but have not yet had any experience with the use of cryptocurrency in general, nor with the use of cryptocurrency via the PayPal app. The reason for this is that the study aims to extend theoretical understanding and explore psychological effects regarding the usage intention of potential users in order to formulate implications and actions for cryptocurrency providers targeting non-users to attract them as future customers. Due to the different levels of experience between users and non-users, different results and therefore implications can be assumed. After this procedure, we obtained a data set of N = 398 random German participants in total, which according to a power analysis in SPSS represents a sufficiently large sample size for the calculation of our model.

The rather uneven gender distribution (58% male) can be explained by the fact that more men tend to use cryptocurrency than women (CoinDesk, 2015). In addition, the age structure of the participants shows particularities. The majority of the participants (93%) are under 30 years of age. This can be explained by the fact that both the online platform and social channel address a predominantly younger target group. Nevertheless, this is important for the further analysis of the data, as especially younger users mainly use cryptocurrencies (Bohr and Bashir, 2014).

Furthermore, we relied on established constructs from previous research and adapted them accordingly for the context of our study. These constructs were first adapted in relation to PayPal and then also adapted in the relation to the use of cryptocurrency via PayPal. To measure perceived trust, a construct was combined using three items of a study by Kim et al. (2011) and two items of Lu et al. (2010) both analyzing the trustworthiness of websites. Our scale to capture perceived risk was based on a study of Featherman and Fuller (2003). Moreover, we adapted a construct of Kim et al. (2011) with four items to measure perceived security. Finally, we adapted the approaches of Chen (2006) and Venkatesh et al. (2012) to measure the intention to use. All items were measured via a seven-point Likert scale (1 = "I totally disagree"; 7 = "I totally agree") and are shown in Table 3.11 including the result of the explorative factor analysis.

Table 3.11 Measurement of the Scales

Variable	Items	Factor Loadings
Perceived Trust in PayPal (α = .92), (Kim et al., 2011; Lu et al., 2010)	PayPal is trustworthy.	.786
	I have confidence in PayPal.	.741
	PayPal is reliable.	.575
	PayPal is fair in its use of private user data.	.594
Perceived Trust in cryptocurrencies when handled by PayPal (α = .95), (Kim et al., 2011; Lu et al., 2010)	Cryptocurrencies are trustworthy if they can be used at PayPal.	.930
	I have confidence in cryptocurrencies when they can be used at PayPal.	.974
	Cryptocurrencies that can be used at PayPal are reliable.	.857
	Payments in cryptocurrencies via PayPal are fair in their use of private user data.	.917

(continued)

Table 3.11 (continued)

Variable	Items	Factor Loadings
Perceived Risk using PayPal (α = .82), (Featherman and Fuller, 2003)	I consider it risky to use PayPal.	.796
	I think it is dangerous to use PayPal.	.739
	PayPal may not work well and may not process payments correctly.	.530
	If transaction errors occur, I am concerned that I may not be able to receive any compensation from PayPal.	.846
Perceived Risk of using cryptocurrencies when handled by PayPal (α = .87), (Featherman and Fuller, 2003)	I think it is risky to use cryptocurrencies if they can be used on PayPal.	.861
	I think it is dangerous to use cryptocurrencies if they can be used on PayPal.	.866
	Cryptocurrencies may not work well at PayPal and payments will not be processed correctly.	.704
	If transaction errors occur, I am afraid that if I use cryptocurrencies, I will not be able to receive compensation from PayPal.	.842
Perceived Security of PayPal (α = .86), (Kim et al., 2011)	Using PayPal is safe.	.663
	Using PayPal is risk-free.	.692
	My privacy is guaranteed with PayPal.	.906
	PayPal protects my data.	.934
Perceived Security of cryptocurrencies when handled by PayPal (α = .88), (Kim et al., 2011)	Using cryptocurrencies with PayPal is safe.	.828
	Using cryptocurrencies at PayPal is risk-free.	.752
	My privacy is guaranteed when using cryptocurrencies at PayPal.	.871
	My data is safe when using cryptocurrencies at PayPal.	.893
Intention to use cryptocurrencies when handled by PayPal (α = .94), (Cheng et al., 2006; Venkatesh et al., 2012)	I would use cryptocurrencies at PayPal.	.957
	I intend to use cryptocurrencies at PayPal for online purchases.	.861

(continued)

Table 3.11 (continued)

Variable	Items	Factor Loadings
	I intend to save money with cryptocurrencies at PayPal.	.751
	I can imagine using cryptocurrencies at PayPal.	.970
Intention to use cryptocurrencies (α = .91), (Cheng et al., 2006; Venkatesh et al., 2012)	I would use cryptocurrencies.	.848
	I intend to use cryptocurrencies for online purchases.	.759
	I intend to save with cryptocurrencies.	.676
	I can imagine using cryptocurrencies.	.976

In addition, we supposed that multicollinearity was not a problem, since variance inflation factors (VIF) were tested and all showed a measurement below the recommended threshold of 10 (Liu et al., 2003). All variables were measured via reflective scales. Also, the measurement models' internal consistencies were evaluated by calculating Cronbach's Alpha, meeting the required cut-off value of at least .7 by all scales (Hair et al., 2014), showing a value of .82 and above for all scales (Table 3.11) and therefore reflecting high levels of scale consistency. In addition, we assessed all scales for discriminant validity by applying Fornell and Larcker's (1981) criterion, indicating that discriminant validity should not be a problem because no scale shares more variance with any other scale than with its own indicators. Furthermore, we conducted the Kaiser-Meyer-Olkin Measure of sampling adequacy, resulting in values of at least .674 and above, which are considered as satisfying as they exceed a value of 0.5. In addition, the total variance explained shows a cumulative value of 72.052%, which also indicates a very good result. Furthermore, using Bartlett's Test of Sphericity, the Chi square values lay at 735.649 and above with significances of < 0.001 and 0.000, also showing satisfying values.

To test our hypotheses, we calculated regression analyses using SPSS 27. Here, the data analysis can be divided into two steps. While in the first step the regressions for the consideration of the transfer of determinants from PayPal to cryptocurrency when handled by PayPal were considered, in a second step the influence of these determinants related to cryptocurrency when handled by PayPal on the intention to use a) cryptocurrency when handled by PayPal and b) cryptocurrency in general was considered.

3.3.4 Results

First, the results of transferring the attributes of PayPal to the use of cryptocurrency via PayPal (the first step of the research model) are presented.

Results show a positive influence of perceived trust in PayPal on trust in cryptocurrency when using PayPal ($F(1, 396) = 17.561$, $p < .001$). The R2 (adjusted R2) of the dependent variable shows a value of .042 (.04), indicating that at least 4% of the variance is explained by perceived trust in PayPal, which according to Cohen (1977) corresponds to a weak to medium effect. The assumption of H1 can be accepted due to the ß-value of .206 and p-value of less than .001. Examining H3, we find that the perceived risk of using PayPal has an impact on the perceived risk of cryptocurrency when using PayPal ($F(1, 396) = 79.628$, $p < .001$). R2 (adjusted R2) is .167 (.165) and H3 is considered accepted (ß = .409, $p < .001$). Results of H5 show a significant impact of the perceived security of PayPal on the perceived security of cryptocurrency when using PayPal ($F(1, 396) = 164.556$, $p < .001$). Moreover, an R2 (adjusted R2) of .294 (.292) was measured. H5 can as well be considered as accepted (ß = .542, $p < .001$).

Next, the effects on the intention to use cryptocurrencies if handled by PayPal as well as in general (the second step of the research model) are presented. Testing for H2a shows that perceived trust in cryptocurrency when using PayPal has a significant effect on the intention to use cryptocurrency via PayPal ($F(1, 396) = 244.488$, $p < .001$). The R2 (adjusted R2) shows a value of .382 (.38). H2a is therefore considered accepted (ß = .618, $p < .001$). Furthermore, the result of H2b shows a significant influence of trust in cryptocurrency when using PayPal on the intention to use cryptocurrency in general ($F(1, 396) = 111.042$, $p < .001$). With an R2 (adjusted R2) of .219 (.217), H2b can also be accepted (ß = .468, $p < .001$). Considering the influence of the perceived risk of cryptocurrency when using PayPal on the intention to use cryptocurrency via PayPal also shows a significant effect ($F(1, 396) = 95.554$, $p < .001$). With an R2 (adjusted R2) of .194 (.192) and a negative influence (ß = −.441, $p < .001$), H4a can be confirmed. Additionally, the results of H4b show an influence of the perceived risk of cryptocurrency when using PayPal on the intention to use cryptocurrency in general ($F(1, 396) = 22.348$, $p < .001$). With an R2 (adjusted R2) of .053 (.051), the hypothesis can be accepted (ß = −.231, $p < .001$). Furthermore, the analysis shows an influence of the perceived security of cryptocurrency when using PayPal on the intention to use cryptocurrency via PayPal ($F(1, 396) =$ 136. 722, $p < .001$). Thus, with an R2 (adjusted R2) of .257 (.255), H6a can be accepted (ß = .507, $p < .001$). Additionally, testing H6b reveals that the perceived security of cryptocurrency when using PayPal has an influence on the intention to use cryptocurrency in general (F(1, 396) = 34.51, $p < .001$). Results show an R2 (adjusted R2) of .08 (.078), confirming H6b (ß = .283, $p < .001$). ß- and F-values are summarized in Table 3.12.

Table 3.12 Results

	Stand. Coef.	F(1,396)
Trust-Transferring Hypotheses		
H1: Perceived trust in PayPal -> Perceived trust in cryptocurrencies if handled by PayPal	.206***	17.561
H3: Perceived risk using PayPal -> Perceived risk of using cryptocurrencies if handled by PayPal	.409***	79.628
H5: Perceived security of PayPal -> Perceived security of cryptocurrencies if handled by PayPal	.542***	164.556
Hypotheses concerning the Intention Behavior		
H2: Perceived trust in cryptocurrencies if handled by PayPal -> Intention to use a) cryptocurrencies if handled by PayPal b) cryptocurrencies in general	a) .618*** b) .468***	244.488 111.042
H4: Perceived risk of using cryptocurrencies if handled by PayPal -> Intention to use a) cryptocurrencies if handled by PayPal b) cryptocurrencies in general	a) −.441*** b) −.231***	95.554 22.348
H6: Perceived security of cryptocurrencies if handled by PayPal -> Intention to use a) cryptocurrencies if handled by PayPal b) cryptocurrencies in general	a) .507*** b) .283***	136. 722 34.51

Note: *significant at $p < .05$; **significant at $p < 0.01$; ***significant at $p < .001$.

3.3.5 Discussion and Conclusion

In our research, we mainly addressed two questions focusing on the transfer of perceived trust, risk and security of PayPal on cryptocurrencies when handled by PayPal as well as the effect of these transferred determinants on the intention to use cryptocurrencies in general. The following discussion and implications for theory and practice are presented based on the two research questions.

3.3.5.1 Implications with Respect to Research Question 1

When considering the first research question, a transfer of all mentioned trust-related attributes from PayPal to cryptocurrency when used via PayPal was expected with respect to the Trust-Transfer Model (Stewart, 2003). Surprisingly, it is not trust that has the strongest effect in the Trust-Transfer Model, but security followed by perceived risk. And while previous research examining the Trust-Transfer Model in the payment sector focuses primarily on the transfer of trust (Gong et al., 2020; Ofori et al., 2017) and numerous research studies concentrate foremost on perceived security and risk as essential determinants for the intention to use technologies and services (Bhatnagar and Ghose, 2004; Tandon et al., 2018), we were also able to show the transfer of security and risk between service provider and payment method. Thus, we provide additional insight into the ways in which attributes such as perceived security, risk, and trust can be transferred, also justifying the consideration of extending the established Trust-Transfer Model to include additional determinants like security and risk. In addition, due to the lack of central regulation and the complex system behind the blockchain (Brezo and Bringas, 2012), it seems that cryptocurrencies are perceived differently by users compared to many other currencies or technologies. As a result, users often have to engage more intensively with the system behind cryptocurrencies before understanding and using them (Arli et al., 2021), which can result in a greater involvement with cryptocurrencies compared to other currencies and technologies. In contrast to the uncertainties when dealing with cryptocurrency, previous studies have found that most users consider PayPal to be the most secure payment method available (Seidenschwarz et al., 2020). The transfers of security and risk from PayPal to cryptocurrencies measured in this study suggest that low security concerns are transferred to cryptocurrencies themselves as well. For cryptocurrency providers, this means that users pay more attention to the perceived security and risk of the provider when trading with cryptocurrency and no additional precautions need to be taken to communicate additional security and reduced risk concerning the trade with cryptocurrency itself.

3.3.5.2 Implications with Respect to Research Question 2

Looking now at the second research question and the influences that the transfer of trust attributes has on user behavior intentions, promising assumptions can be made for cryptocurrency providers here as well. Overall, trust and security of cryptocurrencies when traded via PayPal positively influence the intention to use cryptocurrencies via PayPal. Perceived risk, instead, diminishes this usage intention. While these results primarily confirm past research (Fleischmann and Ivens,

2019; Gil-Cordero et al., 2020; Pavlou, 2003; Rehman et al., 2020), new insight can be obtained for practice Although the preceding attributes are attributed to cryptocurrencies via PayPal, a common increase of the intention to use cryptocurrency in general can be identified. Thus, synergy effects can be observed here, which not only benefit cryptocurrency use via PayPal, but also other cryptocurrency providers and the long-term adoption of cryptocurrency. Here, a general strengthening of the cryptocurrency market might be achieved. One reason for this could be that PayPal, as a brand with a high reputation, serves as a kind of security and trust anchor for cryptocurrency trading (Casalo et al., 2007; Koufaris and Hampton-Sosa, 2004). Users may lose their fear and uncertainty to trade with cryptocurrency if a company/institution they trust and even already use themselves, enters the market. Thus, the perceived attributes of a new payment option (service) via a provider (company) are also transferred to the payment option itself. The prejudices and fear that existed previously may therefore play a lesser role, as users believe that this payment method must generally be secure if a trusted company offers it. Therefore, the payment service provider should pay particular attention to communicate the linkage of their platform and the currency to transfer their attributes and in particular minimize security concerns and perceived risk. With the increasing usability of cryptocurrencies via PayPal, the payment service provider can also benefit. The introduction in the US, for example, has already shown that PayPal users log in more frequently if they also trade cryptocurrencies via the provider (PayPal, 2021a). In addition to implications for practice, new findings for theory and research can also be identified. Results suggest that an extension of the classic Trust-Transfer Model with both trust-related attributes perceived risk and security was successful. With an expansion of the model, new knowledge can be gained about the extent to which users can transfer perception they have towards a (payment) service provider to new services of this payment provider and whether, and if so, which measures are required, to influence the general perception of this service by users. Thus, new theoretical insights could be gained by not only extending the context of the Trust-Transfer Model (originally from company to company or service to service, but here also from company to service), but also by showing that besides trust, other trust-related determinants are essential for behavioral intention forming.

3.3.5.3 Further Research and Limitations

However, for further research it would be interesting to examine whether this extension can also be conducted with other, non-trust-related attributes. Also, although the Trust-Transfer Model was successfully extended in this study, the question arises whether this extension can be generalized and applied to other contexts. Furthermore, the empirical study focused on the most well-known and widely used cryptocurrency, Bitcoin. Since trust, security perception and risk assessment towards the provider are paramount for the intention to use cryptocurrencies, it can be assumed that the usage intention primarily depends on the providers and less on the cryptocurrencies used, which makes it even more important that a platform using cryptocurrencies publicly shows its close connection to this digital currency. But, at this point, the question arises to what extent the original idea of making cryptocurrencies independent of intermediaries still exists today, since service providers such as PayPal increasingly enter the cryptocurrency market. Although, on the one hand, they only offer an additional platform to use cryptocurrencies and there is still no interference from institutions such as banks, on the other hand, market players such as PayPal intervene in the cryptocurrency market and can thus fundamentally influence it, whereby also paving the way for future intermediaries.

In addition, the results of this work are not only interesting for payment service providers who, similar to PayPal, would like to open up to the trading and usage of cryptocurrencies on their platforms, but also for other organizations, networks or companies that offer the handling and trading of cryptocurrency via or on their platforms. For example, there is also an increasing number of companies which accept payments with cryptocurrencies, such as Tesla, or credit cards with which payments with cryptocurrencies are possible. Also, for governments, which offer or would like to offer cryptocurrencies as a national payment method in the future, the findings of this study can help to better assess and analyze usage behavior. Additional research could therefore examine how trust in the government and a country's economic situation might affect cryptocurrency usage. Also, a comparison between countries with strong economies compared to weak economies in terms of their cryptocurrency usage would be an additional way for future research studies to generate new insights that would allow predictions for future cryptocurrency usage.

Nevertheless, the present study also has several limitations. Since literature of trust is incredibly extensive, we tried to cover the main approaches for the context of our study. However, future research might dig deeper into the role of trust in our research context. Further research could investigate whether a transfer of (trust) attributes would also occur with other payment service providers, such as traditional house banks, or other cryptocurrency providers, such as social media platforms since in this study only one specific payment service provider is considered. This research could, thus, focus on the adoption of cryptocurrency by social media platforms, such as currently with Facebook. Additionally, research should be conducted across a broader age structure for examining how the transfer of trust attributes behaves across different age groups. Furthermore, only German-speaking participants were surveyed, who have already used PayPal, but have not yet had any experience with the use of cryptocurrency in general, nor with the use of cryptocurrency via the PayPal app, whose service is so far only possible for the American market. It is therefore not completely sure whether the intention of the non-users ultimately results in actual behavior. Especially for inexperienced non-users, although the same basic knowledge about cryptocurrency was guaranteed at the beginning of the study, their formation of intentions, also for general usage, could be reflected primarily by the perceived trust towards PayPal as one of the largest and most secure online payment service providers. However, we already discussed earlier the important implications that result of the choice of sample (especially for providers trying to reach new users).

3.4 Essay 4—Digital Nudging to Reduce Product Returns in E-Commerce: Investigating Consumers' Intentions and Environmental Implications

This study was published as: Schmitz, J.; Selter, J. L.; Fota, A. & Schramm-Klein, H., 2025. Digital Nudging to Reduce Product Returns in E-Commerce: Investigating Consumers' Intentions and Environmental Implications. Proceedings of the 2025 HICSS.

The essay in this dissertation is partly adapted compared to the published study.

3.4.1 Introduction

The prevalance of online purchases has become firmly established, representing one of the most common means of purchasing products (Kalpoe, 2020). Related to this, online retailers, especially in the fashion industry (Pandey and Chawla, 2018), experience return rates of 50 percent or higher (Asdecker et al., 2016). Returns are a major problem from an economic perspective (Shehu et al., 2020), causing a loss of up to a third of retailers' revenues (Bozzi et al., 2022). Furthermore, they are a major problem from an ecological perspective (Asdecker and Karl, 2022), producing emissions when carried by road, rail, sea, or air and thus foster climate pollution in a sector that generally has a major impact on the environment (Hill and Lee, 2012). However, while returns have economic and environmental impacts, they cannot be banned outright by retail policies. Return options are a crucial driver of consumer loyalty towards retailers and serve as a means of consumer protection (Yang et al., 2017b). The primary reasons for returns are inaccurate product sizing and not fitting clothes (Lee and Xu, 2020). Due to uncertainties about fit, orders with products in different sizes linked with the planned return of the redundant item are becoming more frequent. Therefore, an e-commerce architecture that focuses on "sustainability by design" is necessary (Mirbabaie et al., 2021).

In this context, the scientific discourse focuses on behavioral interventions using nudging elements to guide individuals in the decision-making process in physical (Dolan et al., 2012; Thaler and Sunstein, 2009) and digital environments (Berger et al., 2020; Schrills et al., 2020). Rather than restricting the number of options available, (digital) nudges intend to support better decision-making by modifying the choice architecture: the shape of the context and environment in which consumers' make decisions (Berger et al., 2022). Digital nudges can be understood as an alternative to policies like restrictions, bans or fines (Mirbabaie et al., 2021) as the approach implies an inconspicuous motivation of consumers to behave in ways that seem objectively beneficial to them (Sunstein, 2014). Consequently, digital nudges are to be regarded as behaviorally intervening choice architecture elements, wherein institutions, as decision architects, deliberately "nudge" the decision-making towards the most desirable choice and ultimately influence individuals' behavior (Bucher et al., 2016). Thus, digital nudges can be an effective approach to change consumers' pro-environmental behavior, leading to a reduction of environmental threats (Maki et al., 2016). Digital nudges can address different aspects and are applied in different ways. For example, (Berger et al., 2022) have successfully investigated the impact of digital nudges (*default rules and framing*) on energy-saving behavior in the context of smart home apps. Moreover, Lehner et al. (2016) confirmed digital nudges (*default rules and simplification*) to be effective in encouraging sustainable behavior in energy consumption, personal transportation, and food. Furthermore, the effectiveness of *social norms* (Kroll et al., 2019) and *incentives* (Handgraaf et al., 2013) to foster pro-environmental behavior has also been confirmed. While these studies have demonstrated the influence of digital nudges on pro-environmental behavior, other studies confirmed the effectiveness of digital nudges in the context of customer engagement and app reusage (Wijland et al., 2016) as well as purchase decisions (Zhu et al., 2020).

However, there is a lack of research on digital nudges in the increasingly important context of sustainable (mobile) e-commerce. Previous research suggests that the effectiveness of digital nudges depends on the specific context in which they are applied (Berger et al., 2022). Thus, from both an environmental as well as a managerial perspective, it is interesting to investigate how consumers' behavioral intentions regarding an order including products in different sizes and a planned return change when digital nudges are applied. Next to the missing investigation of digital nudges in the context of consumers' pro-environmental behavior in (mobile) shopping apps studies fall short on investigating the underlying interdependency of digital nudges and the promotion of pro-environmental behavior. Hereby, we intend to fill this research gap by referring to goal framing theory (Lindenberg and Steg, 2007) according to which consumers' goal frames are particularly important for pro-environmental behavior, namely the hedonic goal frame, the normative goal frame and the gain goal frame. Goals are behavior-regulating mental representations of a desired state (Puca and Schüler, 2017) varying from person to person (Westin et al., 2020). Consequently, each goal frame corresponds to a behavioral motivation (Tang et al., 2020). Moreover, individuals' goals are influenced by situational and environmental cues (Lindenberg and Steg, 2007), which can be modified by digital nudges to direct individuals' attention and alter their perceptions (Caraban et al., 2019). This ultimately affects consumers' motivations and corresponding behavioral intentions (Lehner et al., 2016). Consequently, we draw on goal framing theory explaining the effect of the digital nudges on individuals goal motivations and finally on the production of pro-environmental behavioral intentions.

Summing up our practical and theoretical considerations, with this study we seek to analyze the following research questions: Do digital nudges promote consumers' pro-environmental and purchase-usage related behavior in mobile shopping? How is the intention to place an order including products in different sizes and to return the producted influenced by consumers' goal motivations? The present study extends previous research in digital nudge in (mobile) e-commerce as a highly relevant field to information systems by analyzing how digital nudges influence 1) consumers' pro-environmental and 2) purchase-usage related behavioral intentions. We further 3) analyze the influence of consumers' goal motivations on the relationship between the digital nudges and consumers' pro-environmental behavioral intentions.

3.4.2 Theoretical Framework and Hypotheses Developement

With our study we investigate the influence of default, social norm, graphical, incentive, and framing nudges on consumer's pro-environmental, purchase, and usage related behavioral intentions. Our conceptual model builds on the goal framing theory. However, the foucs of this investigation is on exploring the effects of the hedonic, normative, and gain goals explaining the effect of the digital nudges on consumer behavior. The conceptual framework is illustrated in Fig. 3.7.

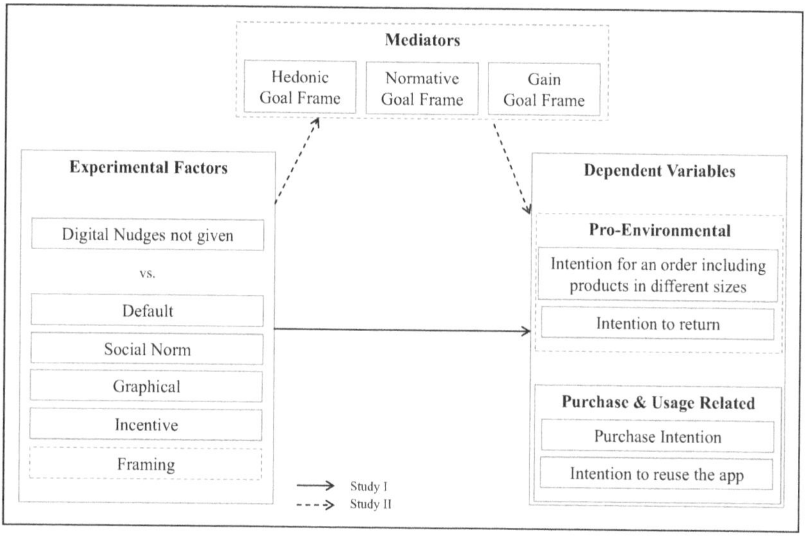

Fig. 3.7 Research Model

3.4.2.1 Digital Nudging

The study of human behavior is complex, as it is influenced by both conscious and unconscious decision-making processes (Evans, 2008). Additionally, various heuristics and biases can impact consumer behavior (Tversky and Kahneman, 1981). In this context, the concept of nudging has emerged as a strategical intervention drawing on behavioral science and aiming to influence the underlying psychological effects so that the decision outcome becomes predictable (Berger et al., 2022). Digital nudges can either be used to reduce possible biases of controlled-considered processes (Hansen and Jespersen, 2013) or to promote automatic-intuitive processing (Mirsch et al., 2017).

However, as consumption decisions are increasingly made in digital (online) environments, Weinmann et al. (2016, p. 433) transferred the approach of nudging to digital environments, defining digital nudging as the "use of user-interface design elements to guide people's behavior in digital choice environments". Consequently, the user-interface represents the digital decision architecture, which is modified by digital nudges in terms of design and content (Tversky and Kahneman, 1981), thereby presenting incentives for user experience (Mirsch et al., 2017). Mirsch et al. (2017), Lehner et al. (2016) and Weinmann et al. (2016) identified several digital nudges, including default rules, social norms, graphical, incentives, and framing nudges, which form the primary object of our research.

Default nudges are characterized by choice architectures where the preferred choice is pre-selected (Thaler and Sunstein, 2009) and thus determines the result if indivuals take no action (Lehner et al., 2016). Moreover, the underlying psychological effect refers to the need to procrastinate due to the time and effort required and preserving the status quo (Berger et al., 2022). The potential of default nudges to influence consumer behavior has been validated across various contexts, such as energy consumption (Hankammer et al., 2021) and environmentally-friendly everyday behavior (Berger et al., 2022). Besides that, default nudges also have shown to positively impact individuals' purchase and usage intention (Hauslbauer et al., 2022). As default nudges have shown to close individuals' gap between attitude and behavior (van Gestel et al., 2021) we argue that pre-selecting the more sustainable choice by default might decrease the number of orders with products in different sizes and returns.

Moreover, social norm nudges aim to encourage certain behaviors by highlighting the actions or choices of others in a way that suggests what is considered typical or desirable within a particular social context. Social norm nudges rely on indiviual's tendency to conform to the perceived norms and standards of their social groups (Mirsch et al., 2017). In this context, Sparkman and Walton (2017) have shown that dynamic information are particularly effective in leading to sustainable behavioral choices. In addition, when norm comparisons consistenly present the same information, Allcott and Kessler (2019) assume a risk of wear and tear. Nevertheless, social norm nudges have been successfully studied in diverse contexts promoting financial (John and Blume, 2018) and energy-saving behavior (Delmas and Lessem, 2014). Hence, highlighting that ordering just one size is considered typical within this specific context could potentially decrease consumers' intentions to place a double order and return.

Furthermore, graphical nudges are characterized by visual clues or graphical elements and aim at the salience bias, referring to indiviual's tendency to perceive salient stimuli particularly strongly (Sunstein, 2014). In this particular context, the significance of colour selection becomes apparent, as colours can have both affective and conative impacts (Lichtlé, 2007). Hereby, warm colours have been shown to have the ability to generate heightened arousal and increased attention (Shapiro and MacInnis, 2002). The employment of graphical nudges to regulate sustainable behavior has been successfully investigated in the context of food consumption (Gottselig et al., 2023). Further, Baltes and Hayward (1976) demonstrated that graphical cues have the potential to encourage greater waste disposal. Thus, we assume that graphical warning concerning the environmental concequences is effective to foster consumers' pro-environmental behavioral intentions.

Additionally, an incentive nudge refers to a type of behavioral intervention that employs positive reinforcements or rewards to influence individuals' decision-making (Hilton et al., 2014) and is based on reciprocity bias (Dur et al., 2010). The effectiveness of incentive nudges to influence pro-environmental behavior has been confirmed, however, a certain strength of incentive is required in the first place to activate the controlled-considered system. Furthermore, the impact can diminish once the incentive is removed (Handgraaf et al., 2013). Hence, rewarding the decision against placing an order including products in different sizes and return might be helpful to increase consumers' pro-environmental, purchase and usage related behavioral intentions.

Framing nudges aim at transporting condensed information about complex subjects or highlighting specific characteristics more noticeably (Berger et al., 2022). In addition, the automatic and intuitive tendencies of human behavior cause individuals to take notice only of information being relevant to them personally indicating that consumers frequently overlook crucial environmental indicators whilst making daily purchase decisions indicating an attitude-intention gap (Allcott and Taubinsky, 2015). Furthermore, studies indicate that emphasizing environmental impacts strengthens consumers' awareness and sense of responsibility for sustainable consumption (Tang et al., 2020) as well as the purchase intention (Zhu et al., 2020). Consequently, framing nudges revealed promising results for promoting sustainable decisions in the context of environmental-friendly behavior (Berger et al., 2022). Hence, breaking down the complex information concerning the environmental concequences when placing an order including products in different sizes and conducting returns might be helpful to decrease consumers' intentions for such actions. In summary, we expect that digital nudges will have a negative impact on consumers' intention to place an order including products in different sizes and consumers' intention to return. In contrast, digital nudges will positively influence consumers' purchase and usage related behavioral intentions.

H1: The digital nudges a) default, b) social norm, c) graphical, d) incentive, and e) framing have a negative effect on consumers' intention of conducting an order including products in different sizes.

H2: The digital nudges a) default, b) social norm, c) graphical, d) incentive, and e) framing have a negative effect on consumers' intention of returning products.

H3: The digital nudges a) default, b) social norm, c) graphical, d) incentive, and e) framing have a negative effect on consumers' purchase intention.

H4: The digital nudges a) default, b) social norm, c) graphical, d) incentive, and e) framing have a negative effect on consumers' intention to reuse the app.

3.4.2.2 Goal Framing

Goal framing theory was established by Lindenberg and Steg (2007) and states that indiviuals' behavior is driven by different goals. The authors posit that goals regulate an individual's focus, determine which information and perspectives are most accessible, influence evaluations of various aspects of a situation, and impact the consideration of alternatives (Lindenberg and Steg, 2007). Goals are mental representations of a desired statethat regulate behavior (Puca and Schüler, 2017). Each goal corresponds to a behavioral motivation (Tang et al., 2020). According to this theory, individuals' goals are activated situational and environmental cues (Lindenberg and Steg, 2007). Digital nudges can enhance or modify this activation by manipulating the salience to guide attention and alter perceptions (Caraban et al., 2019), thereby influencing motivations and corresponding behavioral intentions (Chen et al., 2016; Lehner et al., 2016). Consequently, we draw on Lindenberg and Steg's (2007) goal framing theory to explain the effect of digital nudges on individuals goal motivations and finally on the production of individuals' pro-environmental behavioral intention (Yang et al., 2020).

Lindenberg and Steg (2007) identified three goal frames being particularly important for behavioral change, namely the hedonic goal frame, the normative goal frame, and the gain goal frame. The hedonic gain goal refers to individuals striving for positive feelings achieved in the process and outcomes of specific behaviors (Steg et al., 2014). According to Lindenberg and Steg (2007) the hedonic goal frame activates goals aiming to improve how individuals feel in a particular situation, to entail seeking for pleasure and to avoid negative feelings (Lindenberg and Steg, 2007). In this context, several studies have shown, that anticipation of future emotions can be regarded as a primary motivation of behavior (Rezvani et al., 2018), especially regarding drivers of pro-environmental behavior (Bamberg and Möser, 2007). In line to Steg and Vlek (2009), we assume that framing nudges concerning the environmental concequences of an order including products in different sizes and return activate consumers' hedonic goal frame, as they nudge consumers' negative feelings like guilt and discomfort (Connolly and Zeelenberg, 2002), resulting in pro-environmental behavioral intentions:

H5a: The influence of framing on consumers' intention of conducting an order including products in different sizes is mediated by the hedonic goal frame.

H6a: The influence of framing on consumers' intention of returning products is mediated by the hedonic goal frame.

However, the normative goal frame is associated to all types of goals regarded as appropriate, such as behaving properly, showing exemplary behavior, and contributing to a clean environment (Mirsch et al., 2017). The goals in question pertain to perceived norms regarding how one should behave (Cialdini et al., 1990). This prompts individuals to consciously seek behaviors that are perceived as appropriate for personal norms (Yang et al., 2020). Studies have shown that consumers' personal moral norms and the resulting normative motivations are predictors for consumers' pro-environmental behavior (Jansson et al., 2017). In this context, Groot and Steg (2009) identified consumers' environmental knowledge, awareness of consequences, and assumption of responsibility as decisive factors. Based on this, we expect framing the environmental concequences to enhance consumers' awareness for environmental consequences. Furthermore, this encourages personal responsibility and thus activates the normative goal frame, influencing consumers' pro environmental behaviors (Westin et al., 2020). Based on the literature, following hypotheses are proposed:

H5b: The influence of framing on consumers' intention of conducting an order including products in different sizes is mediated by the normative goal frame.

H6b: The influence of framing on consumers' intention of returning products is mediated by the normative goal frame.

The gain goal frame makes individuals aware of changes in their personal resources, including both materialistic and non-materialistic resources (Donmez-Turan and Kiliclar, 2021). Accordingly, to Noppers et al. (2014) and Griskevicius et al. (2010) consumers' intentions towards pro-environmental behavior increase when people believe their actions lead to enhancing their status. Accordingly, following hypotheses are proposed:

H5c: The influence of framing on consumers' intention of conducting an order including products in different sizes is mediated by the gain goal frame.

H6c: The influence of framing on consumers' intention of returning products is mediated by the gain goal frame.

Overall, we argue that framing for environmental consequences of order including products in different sizes and the corresponding returns as a deliberate description and simplification modifyies and enhances the virtual decision architecture (Mirsch et al., 2017) and thus consumers' situational cues (Caraban et al., 2019). As a result, consumers' goal frames and the corresponding behavioral intentions are activated. Moreover, we assume that this strategic change of the digital choice architecture raises consumers' awareness to the environmental concequences, establishing a cognitive link between concrete behaviors (order including products in different sizes and the resulting return) and environmental consequences. As the goal framing theory states that all three goals can be activated at the same time (Lindenberg and Steg, 2007), we hereby consider consumers' hedonic goal as their attitude towards placing an order including products in different sizes and returning. Furthermore, the normative goal is defined with respect to consumers' environmentalism and the gain goal in form of social status as predictors of consumers' pro-environmental behavioral intentions.

3.4.3 Study 1

3.4.3.1 Empirical Study: Method and Procedure

To analyze the influence of different forms of nudges as app design elements, we conducted an online experiment, using a between-subject design with six scenarios. All subjects were told to put themselves in the process of purchasing a gender-neutral pair of shoes online via a mobile app. Hereby, a fictional, app-based online store and the corresponding shopping cart were presented. Due to uncertainties about size caused by the lack of a physical product experience, all subjects were asked to imagine the same product being in two different sizes in their shopping cart. Furthermore, subjects were shown the check-out process of a potentially occurring order including products in different sizes. Based on Weinmann et al. (2016), we integrated default, social norm, graphical, incentive, and framing nudges as follows: In the control scenario, the shopping cart contained shoes along with information indicating, that participants had placed the same shoe in two different sizes in their shopping cart. The default nudge ist characterizes by choice options and a sustainable pre-selected option, requiring consumers to manually change their choice and thus encouraging consumers to decide for one size. In the third scenario we implemented a social norm nudge in the app design by providing insights into the decision-making of other users and explicitly highlighted the information that 90 percent of users decided against an order including products in different sizes in the past hour. In szenario four we depicted a red warning triangle as a graphical nudge. In the fifth scenario the incentive nudge is represented by a monetary incentive strategically designed to reward the decision against a double order and dissuade opportunistic returns by offering complimentary shipping. Within the framing scenario we highlighted information about the potentially occurring order including products in different sizes and the resulting return by emphasizing certain aspects. In detail, subjects were told that 765,000 returns arise per day in Germany, resulting in yearly emissions of 238,000 tons of CO_2. This was further illustrated as being equivalent to the distance covered by 6.6 million cars driving from Hamburg to Munich. The scenarios are illustrated in Fig. 3.8.

Fig. 3.8 Experimental Design Study 1

Regarding the dataset, it was ensured that participants were equally divided between the different digital nudges (no digital nudge = 42 vs. default = 65 vs. social norm = 66 vs. graphical = 63 vs. incentive = 61 vs. framing = 55). A total of 352 participants (women: 59.7 %) with an average age of 28.76

years (SD = 8.66) participated in the study. To test the hypotheses, established constructs were adapted to this study and measured on a seven-point-Likert-scale. Furthermore, the constructs were reverse coded if necessary and calculated by mean-aggregation. These constructs are illustrated in Table 3.13. Reliability was measured by calculating Cronbach's Alpha (α) meeting the required value ($\alpha \geq .7$) for all constructs (Hair et al., 2014).

Table 3.13 Constructs Study 1

Construct	Item		α	Abbreviation
Intention to place an order including products in different sizes (Bansal et al., 2004)	How would you assess your intention to order the same product in different sizes under the given given conditions and information?	unlikely / likely	.964	SOI
		improbable / probable		
		no chance / certain		
		probably not / probably		
Intention to return (Burton et al., 1999)	Given the information shown, how likely are you to return the order?		.900	RI
	Given the information shown, how likely is it that you would consider returning the order?			
Purchase Intention (Wang et al., 2011)	After using the app, I intend to purchase from the app immediately.		.868	PI
	After using the app, the likelihood of me purchasing from the app immediately is…			
	I rate my chances of purchasing from the app immediately as…			
Intention to Reuse the shopping app (Pavlou, 2003)	I intend to use this app in the future.		.942	IRA
	I should use this app in the future.			
	I will buy something via this app in the near future.			
Choice Difficulty (Laroche et al., 2003)	Ease of choosing whether to order both sizes or just one		.918	CD
	Choice of whether ordering the same product in two different sizes or not is not very complicated			
	Ease of discriminating among two different sizes of one product			

ANOVA tests were calculated, confirming a successful manipulation of the level of choice difficulty whether to order both sizes or just one within the experiment (CD: no digital nudge = 4,24 vs. default = 3.21 vs. social norm = 3.09 vs. graphical = 3.06 vs. incentive = 3.07 vs. framing = 2.34) $F(1.352) = 8.23$ $p < .001$, $\eta 2 = 0.11$). Further, a Tukey post-hoc test also confirmed significant differences of the level of choice difficulty between the control scenario and the digital nudges. No distortions were found for technology affinity or online-shopping experience between the scenarios. The gender of the survey participants as well as the perceived choice freedom also have no influencing effect.

3.4.3.2 Findings

To measure the influence of the different types of digital nudges on consumers' pro-environmental, purchase and usage related behavioral intentions, ANOVAs were calculated. Following results emerge regarding consumers' pro-environmental behavioral intentions and are also illustrated in Table 3.14. Intention to place an order including products in different sizes:$_{digital\ nudge\_not\_given}$ = 4.99 (1.32), $M_{default}$ = 3.94 (1.59), M_{social} = 3.53 (1.44), $M_{graphical}$ = 3.70 (1.75), $M_{incentive}$ = 3.81 (1.42), $M_{framing}$ = 2.69 (1.63), $F(1, 352) = 10.38$, $p < 0.01$, $\eta^2 = 0.13$; Intention to return: = 4.85 (1.56), $M_{default}$ = 4.01 (1.45), M_{social} = 3.77 (1.70), $M_{graphical}$ = 3.95 (1.56), $M_{incentive}$ = 3.94 (1.48), $M_{framing}$ = 2.93 (1.49), $F(1, 352) = 8.25$, $p < 0.01$, $\eta^2 = 0.11$. Variance homogeneity was ensured, allowing for a Tukey post-hoc test to be carried out. Significant differences in consumers' intention to place an order including products in different sizes and to return were measured between the scenarios. Results indicated that all digital nudges differed significantly from the scenario in which no digital nudge was given. Moreover, the results show a significant dissimilarity in the case of framing in relation to the dependent variables when compared to all other digital nudges. Furthermore, linear regressions confirmed negative effects of the digital nudges on the dependent variables lowering consumers' intention to place an order including products in different sizes ($ß = -0.31$, $p < 0.01$) as well as to return ($ß = -0.28$, $p < 0.01$) significantly, leading to a confirmation of H1a–e and H2a–e.

Table 3.14 Results Study 1

Dependent Variable	Digital Nudges						F (η^2)
	Mean (SD)						
	Not given	Default	Social Norm	Graphical	Incentive	Framing	
Intention to place an order including products in different sizes	4.99 (1.32)	3.94 (1.59)	3.53 (1.44)	3.70 (1.75)	3.81 (1.42)	2.69 (1.63)	10.38*** (0.13)
Intention to Return	4.85 (1.56)	4.01 (1.45)	3.77 (1.70)	3.95 (1.56)	3.94 (1.48)	2.93 (1.49)	8.25*** (0.11)
Purchase Intention	4.01 (1.08)	4.07 (1.31)	3.30 (1.60)	4.11 (1.28)	4.13 (1.34)	4.07 (1.51)	3.49** (0.05)
Intention to Reuse the App	3.68 (1.65)	4.43 (1.59)	3.67 (1.61)	3.24 (1.75)	4.62 (1.36)	4.65 (1.47)	8.51*** (0.12)

** Sig.: p < .05; ** < 0.01; *** p < .001*

Results of the ANOVAs generally support our assumptions of H3 and H4, as digital nudges interact significantly with consumers' purchase and usage related behavioral intentions (Purchase intention:$_{digital\ nudge\_not\_given}$ = 4.01 (1.08), $M_{default}$ = 4.07 (1.31), $M_{social\_}$ = 3.30 (1.60), $M_{graphical}$ = 4.11 (1.28), $M_{incentive}$ = 4.13 (1.34), $M_{framing}$ = 4.07 (1.51), $F(1, 352) = 3.49$, $p < 0.05$, $\eta^2 = 0.05$; Intention to reuse the app: = 3.68 (1.65), $M_{default}$ = 4.43 (1.59), $M_{social\_}$ = 3.67 (1.61), $M_{graphical}$ = 3.24 (1.75), $M_{incentive}$ = 4.62 (1.36), $M_{framing}$ = 4.65 (1.47), $F(1, 352) = 8.51$, $p < 0.01$, $\eta^2 = 0.12$). Again, variance homogeneity was ensured and the results of the Tukey post-hoc test show significant differences in consumers' purchase intention as well as consumers' intention to reuse the app between the scenarios. In detail, results indicated that the social norm nudge differed significantly from all other digital nudges in terms of consumer' purchase intention. However, no significant difference was observed between the default, graphical, incentive, and framing nudge, and the control scenario. Regarding consumers' intention to reuse the app results show significant dissimilarities in the case of incentive, and framing nudges compared to all other scenarios. No significant difference was observed for default, social norm, and graphical nudges compared to the control scenario. Nevertheless, there was a trend showing a decrease of consumers' intention to reuse the app with graphical nudges and an increase with default nudges. Furthermore, linear regressions confirmed a negative significant effect of the social norm nudge on consumers' intention to purchase ($ß = -0.35$, $p < 0.01$). Positive significant effects for the incentive, and framing nudges on consumers' intention to reuse the app ($ß = 0.21$, $p < 0.01$) were measured. Results lead to a confirmation of H3b and H4d and e.

3.4.3.3 Discussion

The study's findings demonstrate that integrating digital nudges into user-interface design significantly impacts consumers' decision-making when initiating orders with products in different sizes, which in turn influences pro-environmental, purchase, and usage-related behaviors. Digital nudges seem to enhance consumers' decision making by reducing the perceived choice difficulty through graphic and textual clarification of specific aspects in the decision architecture (Lehner et al., 2016). Thus, individuals are guided and enabled in the decision-making process (Thaler and Sunstein, 2009), ultimately influencing consumers behavioral intentions (Hummel and Maedche, 2019). Our study confirms positive influences of digital nudges on cosumers' intention to place an order including products in different sizes and to return, both being positively linked to pro-environmental patterns. Moreover, framing appears to be the most efficacious method of modifying decision architecture by simplifying the information provided to encourage consumers' pro-environmental behavioral intentions, supporting findings of Lehner et al. (2016). However, the effect of digital nudges on consumers' purchase and usage related behavioral intentions deviate from those on consumers' pro-environmental behavioral intentions. While all digital nudges, and especially the framing nudge, have a positive impact regarding pro-environmental behavior by lowering consumers' intention to place a seletion order and to return, the social norm nudge seems to have a negative influence on consumers' purchase intention. Regarding consumers' intention to reuse the app, contradictory results were obtained. While the graphic nudge lowers the usage intention, the incentive and especially the framing nudge increase consumers' intention to reuse the app. Overall, results go in line with findings of Berger et al. (2022), Mirsch et al. (2017) and Weinmann et al. (2016) confirming the effectiveness of digital nudges. Findings have shown default, social norm, graphical, incentive and framing nudges to positively influencing consumers' behavioral intentions of acting in a pro-environmental way (Schacker and Stanoevska-Slabeva, 2021). Thus, nudges can serve as an alternative to policy design, such as implementing restrictions, bans or, fines (Mirbabaie et al., 2021).

3.4.4 Study 2

3.4.4.1 Empirical Study: Method and Procedure

Study 1 demonstrated the influence of digital nudges as app design elements and identified framing for the environmental consequences of an order including products in different sizes having the strongest impact on consumers' pro-environmental behavioral intention. Study 2 examines framing for environmental consequences of an order including products in different sizes in more detail, conducting an online expermiment, using a between-subject design with two scenarios. All participants were instructed to imagine purchasing a gender-neutral pair of shoues through the fictional online store used in Study 1. They were asked to envision the same product in two different sizes in their shopping cart. To manipulate consumers' awareness of the environmental impact of their potential order including products in different sizes, we utilized scenario techniques. We presented two scenarios: one with no digital nudge and one that framed the potential return result as having an environmental cost. For the sake of comparability, we adopted the framing approach from Study 1. The two scenarios are illustrated in Fig. 3.9.

Fig. 3.9 Experimental Design Study 2

Respondents were randomly assigned to one of the two experimental conditions (Framing nudge not given = 129 vs. given = 117). A total of 246 subjects (women: 70%) with an average age of 27.91 years (SD = 10.03) took part in the study. To test the hypotheses, established constructs were adapted to this study, all measured on a seven-point-Likert-scale ranging from "not apply at all" (1) to "fully apply" (7) and calculated by mean-aggregation. These constructs are illustrated in Table 3.15. Reliability was measured by calculating Cronbach's

Alpha (α). The required value ($\alpha \geq .7$) was achieved for all measurements (Hair et al., 2014). For the constructs behavioral intention towards an order including products in different sizes ($\alpha = .964$) and intention to return ($\alpha = .90$) the same sources and items were chosen as in study one (see Table 3.13) to ensure comparability. For the experimental design, subjects' evaluations of the perceived environmental cognitive resource demand were used as a manipulation check. ANOVAs were calculated to check whether the manipulation was successful (environmental cognitive resource demand: $M_{framing\_not\_given} = 5.51$ (1.18), $M_{framing\_given} = 4.12(1.50)$, $F(1, 244) = 4.82$, $p < .05$, $\eta 2 = 0.019$). Results confirm a successful manipulation of consumers' awareness for the environmental consequences due to framing for the environmental cost associated with the potential order including products in different sizes.

Table 3.15 Constructs Study 2

Construct (source)	Item	α	Goal Frame
Attitude towards the act (Klein et al., 2004)	I would feel uncomfortable if I bought the same product in two different sizes.	.850	hedonic goal frame
	I would feel guilty if I bought the same product in two different sizes.		
	I feel better about myself if I do not order the same product in two differenz sizes.		
Environmentalism (Schuhwerk and Lefkoff-Hagius, 1995)	My actions, like ordering the same product in two different sizes or return an order, have an impact on the environment.	.837	normative goal frame
	I am concerned about the environment.		
	The condition of the environment affects the quality of my life.		
	I am willing to make sacrifices to protect the environment.		
Value (Sweeney and Soutar, 2001)	Ordering just one size would make good impressions on other people.	.879	gain goal frame
	Ordering just one size would help me to feel acceptable.		

(continued)

Table 3.15 (continued)

Construct (source)	Item	α	Goal Frame
	Ordering just one size would improve the way I am perceived.		
Cognitive resource demand (Lee and Aaker, 2004)	It is easy for me to deal with the advantages and disadvantages of this potential order including products in different sizes and the associated environmental costs.	.877	Cognitive resource demand
	It is easy for me to understand the advantages and disadvantages of this potential order including products in different sizes and the associated environmental costs.		
	It is easy for me to realise the advantages and disadvantages of this potential order including products in different sizes and the associated environmental costs.		

No distortions were found for technology affinity or experience with the use of online stores between the scenarios. Finally, we controlled for the purchase intention ($F(1,244) = 0.037$, $p = 0.847$) and shopping experience ($F(1, 244) = 1.422$, $p = 0.28$) of the respondents, not identifying any systematic distortion between the treatments.

3.4.4.2 Findings

To test the postulated mediating influence of consumers' hedonic, normative and gain goal frames on the effects of the experimental factor on the dependent variables, PROCESS (model 4) was used as suggested by Zhao et al. (2010). Following the approach of Baron and Kenny (1986) and their first of four steps of mediation, significant values for the direct paths of the experimental factor on the dependent variables (consumers' intention to place an order including products in different sizes ($ß = -.626$, $t(244) = -3.13$, $p = .002$) and intention to return ($ß = -.719$, $t(244) = -2.94$, $p = .003$)) could be measured. Furthermore, significant paths for the framing nudge on the hedonic goal frame, normative goal frame and gain goal frame as well as significant paths from the hedonic goal frame, normative goal frame, and gain goal frame on the intention to place an order including products in different sizes and the intention to return could be measured, confirming step two and three of Baron and Kenny (1986). Split results emerge for step four. Regarding the mediating effect of hedonic goal frame, only the indirect path of the framing nudge on consumers' intention to place an order including products in different sizes remains significant, indicating full mediation (relative total effect: ß = −.63, p = .002; relative direct effect: ß = −.34, p = .086). The relative indirect effect shows a significant mediation (relative indirect effect = −.29, SE = .09, 95% CI[−.49, −.13]). Results confirm hypothesis H5a. Moreover, regarding the mediating effect of the normative goal frame, both the indirect and direct path of the framing nudge on the intention to place an order including products in different sizes are significant, indicating partial mediation (relative total effect: ß = −.63, p < .001; relative direct effect: ß = −.51, p = .002). The relative indirect shows a significant mediation (relative indirect effect = −.14, SE = .21, 95% CI[−.28, −.02]). Results confirm hypotheses 5b. Results for the mediating effect of the gain goal frame show significant indirect and direct paths of the framing nudge on the intention to place an order including products in different sizes, indicating partial mediation (relative total effect: ß = −.63, p = .002; relative direct effect: ß = −.53, p = .008). The relative indirect effect shows a significant mediation (relative indirect effect = −.22, SE = .08, 95% CI[−.39, −.08]). Results confirm hypothesis 5c. However, for the results of step four show different results when considering consumers' return intention. Regarding the mediating effect of the hedonic goal frame, both the indirect and direct paths of the framing nudge on consumers' intention to return are significant, thus indicating partial mediation (relative total effect: ß = −.72, p = .004; relative direct effect: ß = −.52, p = .039). The relative indirect effect shows a significant mediation (relative indirect effect = −.21, SE = .09, 95% CI[−.39, −.055]). Results confirm hypothesis H6a. Regarding the mediating effect of the normative goal frame, only the indirect effect of the framing nudge on the intention to return is significant, indicating a full mediation (relative total effect: ß = −.72, p = .04; relative direct effect: ß = −.47, p = .061). The relative indirect effect shows a significant mediation (relative indirect effect = −.25, SE = .084, 95% CI[−.44, −.10]). Results confirm hypothesis 6b. Regarding the mediating effect of the gain goal frame, only the direct path of the framing nudge on the intention to return is significant, indicating no mediation and leading to a rejection of hypothesis 6c.

3.4.4.3 Discussion

Results indicate that retailers framing for the environmental consequences associated with an order including products in different sizes and the return can lower the cognitive effort required to understand the underlying environmental impact. This can increase consumers' environmental awareness and thus influence their pro-environmental behavioral intentions. Consumers seem to perceive the framing nudge as a deliberate formulation, description, and simplification of the virtual decision architecture (Mirsch et al., 2017). Results go in line with research of Thaler and Sunstein (2009) that nudges influence consumer behavioral patterns effectively. In detail, results correspond to the initial study with framing significantly decreasing consumers' intention to place an order including products in different sizes and to return. Moreover, regarding the goal framing theory of Lindenberg and Steg (2007), results indicate hedonic, normative, and gain goal frames being crucial for consumers' pro-environmental behavior. It was shown that the framing nudge activates consumers hedonic, normative, and gain goal motivations, which in turn influence consumers' intention to place an order including products in different sizes and to return. Results go in line with previous literature (Rezvani et al., 2018) stating that consumers' hedonic and normative goal frames can be important mediating factors (focal goals) for consumers' pro-environmental behavioral intentions. However, consumers' gain goals only partially mediate the effect on consumers' intention to place an order including products in different sizes (Yang et al., 2020). The relation between the framing nudge and the intention to place an order including products in different sizes is primarily mediated by consumers' hedonic goal frame. Consumers seem to associate the act of placing an order including products in different sizes predominantly with negative feelings like guilt or discomfort, leading to a significant decrease of the intention to place an order including products in different sizes through the hedonic goal frame. However, regarding the intention to return, the results suggest that consumers' personal moral norms (Jansson et al., 2017) play a decisive role in activating their normative motivation to take personal responsibility, explaining the decrease in the intention to return. Surprisingly, consumers' gain goals have shown no mediating influence in this context. The framing nudge seems to enhance consumers' awareness for environmental consequences and thus activates consumers' motivational goals to prevent negative feelings such as guilt or discomfort.

3.4.5 General Discussion and Implications

Table 3.16 provides an overview of the acceptance or rejection of the postulated hypotheses. Overall, most hypotheses could be supported, and various theoretical approaches can be derived from the results of the empirical studies.

Table 3.16 Summary of Results

Study I	DV	Digital Nudges (IV)				
		Default	Social Norm	Graphical	Incentive	Framing
	Intention to place an order including products in different sizes	confirmed	confirmed	confirmed	confirmed	confirmed
	Intention to Return	confirmed	confirmed	confirmed	confirmed	confirmed
	Purchase Intention	rejected	confirmed	rejected	rejected	rejected
	Intention to Reuse the App	rejected	rejected	rejected	confirmed	confirmed
Study II	IV	Med.	DV	H	Res.	Type of Med.
	FRAMING	hedonic goal frame	Intention to place an order including products in different sizes	H5a	confirmed	full
		normative goal frame	Intention to place an order including products in different sizes	H5b	confirmed	partially
		gain goal frame	Intention to place an order including products in different sizes	H5c	confirmed	partially
	FRAMING	hedonic goal frame	Intention to Return	H6a	confirmed	partially
		normative goal frame	Intention to Return	H6b	confirmed	full
		gain goal frame	Intention to Return	H6c	rejected	–

DV = dependent variable, IV = independent variable, Med. = Mediation, H = Hypothesis, Res. = Results

First, the approach of nudging (Thaler and Sunstein, 2009) in digital environments (Weinmann et al., 2016) was confirmed. Accordingly, consumers' intention to act in a pro-environmental way is higher when digital nudges are present. Consumers seem to perceive such strategic modifications of the user-interface in decision architecture as guidiance (Bucher et al., 2016), encouraging their behavior without mandating or instructing (Halpern, 2015). Based on this, digital nudges represent an alternative to policies like restrictions, bans or fines (Mirbabaie et al., 2021). However, not all digital nudges have shown to be equally suitable to close individuals' gaps between attitude and behavior. While the framing nudge appears to be the most promising method to effectively influence consumers pro-environmental, purchase and usage related behavioral intentions (Berger et al., 2022), the social norm nudge even reduced consumers' intention to purchase and to reuse the mobile app. Overall, results expand the findings of Berger et al. (2022), Mirsch et al. (2017) and Weinmann et al. (2016) not only by confirming the effectiveness of default, social norm, graphical, incentive, and framing nudges all in one study to influence consumers' behavioral intentions, but rather shed light on the psychological mechanisms underlying digital nudging. In this context, the goal framing theory (Lindenberg and Steg, 2007) has been confirmed. Accordingly, we were able to show that consumers' behavioral intentions can be attributed to hedonic, normative, and gain goal frames which explain the effect of the digital nudges on individuals goal motivations, being crucial for consumers' pro-environmental behavior (Westin et al., 2020). Full mediations were supported for the effect of the digital nudge on consumers' intention to place an order including products in different sizes and consumers' intention to return. Hereby, previous literature in the context of private mobility (Westin et al., 2020) and pro-environmental behavior (Groot and Steg, 2009) could be confirmed. Interestingly, consumers seem to attach greater importance to their inner moral and ethical code when making decision in the context of pro-environmental behavior regarding mobile shopping, instead of being influenced by opinions of their social environment. Futhermore, contrary to our expectations, no mediating effect of the gain goal frame on consumers' return intention was measured. Accordingly, literature from the context of sustainable fashion consumption (Lundblad and Davies, 2016) could not be transferred either. Reasons for this could be that consumers automatically tend to associate an order including products in different sizes with an unavoidable return, since both items are interchangeable, which in turn makes the ecological consequences appear more prominent. Conversely, returns themselves are a common practice that addresses the issue of physical product absence in the context of online shopping, and are considered socially and ecologically more appropriate than orders with products in different sizes, particularly for consumers living in less urban areas.

3.4.5.1 Practical Implications

The findings of this study offer valuable practical implications. Retailers, particularly in the fashion industry, face high return rates with negative economic and ecological consequences. However, previous approaches were either ineffective, counterproductive, or intervened in the process after the transaction. Our results imply that digital nudges can be effectively integrated into mobile apps' user-interfaces, proactively mitigating potential orders with products in different sizes and returns. Accordingly, retailers can implent digital nudges, such as default, social norm, graphical, incentive, and framing, to enhance the dicision making of their customers by guiding and encouraging their behavior (Bucher et al., 2016). Hereby, retailers do not mandate or instruct their customers (Halpern, 2015), but reduce the probability of an order including products in different sizes and the number of returns. Thus, the concept of digital nudges pertains to a user interface architecture that focusses on "sustainability by design" (Mirbabaie et al., 2021) to foster customers' pro-environmental behavioral intentions and to counteract environmental threats (Maki et al., 2016). Moreover, by strategically incorporating digital nudges, retailers can enhance user experience on digital platforms, making interactions more user-friendly, engaging, and conducive to positive outcomes (Weinmann et al., 2016). By doing so, retailers can increase customers' intention to reuse the app. Furthermore, especially as strict return polices have been identified to reduce consumer purchase-related behavior (Bower and Maxham, 2012), various digital nudges have not shown these results.

Although digital nudges can contribute to a more sustainable environment by reducing the likelihood of returns (Asdecker and Karl, 2022), and sustainability is a key priority for retailers (Shen et al., 2017), the literature has shown that consumers' primary goal in using mobile shopping apps is to access products and services conveniently, efficiently, and without risk (Natarajan et al., 2018). Thus, retailers should primarily preserve user autonomy and concentrate on the balance between influencing behavior and respecting consumers choices. Accordingly, it should be particularly emphasized that none of the digital nudges investigated, restricted consumers' perceived freedom of choice. However, digital nudges and especially framing should be used according to ethical guidelines and principles and not be abused to manipulate consumers' behavior. Representing a dark pattern of marketing, retailers could frame consumers behavior in contrast to their actual attitude. Overall, the study provides valuable guidance for improved user-interface design by showing retailers wich digital nudges should be payed particular attention for promoting sustainability without restricting consumers and without negatively affecting consumers' purchase and usage related behavior.

3.4.5.2 Limitations and Future Research

In addition to the important theoretical and practical implications of the studies, several limitations must be considered. The studies were conducted as online experiments, possibly biasing results of the study. Therefore, further experiments should examine the results and address the challenges faced due to the applied experimental approach. In addition, the present study examined behavioral intention rather than actual behavior, and it would have been valuable to include behavioral measures as people tend to overreport self-reported behavior due to social desirability bias. Therefore, it is important for future research to also examine the influence of digital nudges on actual consumer behavior. Nevertheless, most studies testing digital nudges and goal framing included only self-reported behavior (Westin et al., 2020) and research has indicated that intentions are closely linked to behavior (Armitage and Conner, 2001). Thus, interesting results could be shown in the studies, also pointing out several opportunities for further research. Since we only focused on each digital nudge individually, and studies have already shown that the combination of behavioral interventions can increase the impact on behavioral intentions (Kroll et al., 2019), further research should explore the combined use of the digital nudges studied. Hereby, research should focus on examining the combination to amplify existing effects or whether they counteract each other. Furthermore, the use of digital nudges raises important ethical considerations regarding user autonomy and privacy. Future research should concentrate on the optimal balance between influencing behavior and respecting individual choices. Finally, the study was conducted in Germany and cross-cultural differences, for example, in terms of personal characteristics and demographics as well as differences in consumer behavior which are decisive factors for the effectiveness of digital nudges (Kroll et al., 2019) might have occured. Accordingly, examining these aspects regarding the effectiveness of digital nudges could yield interesting results.

3.5 Essay 5—Decision Aids for Online Size Selection: Insights from Experimental Studies in Apparel Retailing

This study was submitted by Selter, J. L.; Schmitz, J.; Röding, T. & Schramm-Klein, H., to The International Review of Retail, Distribution and Consumer Research and is currently in the second round of peer review (17th April 2025).

3.5.1 Introduction

Online shopping offers several benefits for customers like a convenient way of shopping, a comprehensive product selection and a lot of information about the individual products available (Blut et al., 2023; Gallin and Portes, 2024) and has increased sharply in the past (Asdecker and Karl, 2022; Statista, 2023). However, online retailing goes along with returns of products. Around 16.5 percent of products are returned in the United States with a value of around $ 210 billion (Guha, 2023).

One of the most sought-after categories in online retailing is apparel (Mendez, 2024). With package-related return rates of 47 to 50 percent in footwear and apparel (Asdecker and Karl, 2022), this category causes the highest return rates (Cullinane et al., 2019). Returns are a major problem from an economic perspective (Hong and Pavlou, 2014; Shehu et al., 2020) and lead to a loss of up to a third of retailers' revenues (Bozzi et al., 2022). In addition, returns may be a troublesome issue for customers, as returns, refunds and new orders require time, costs and effort (Hong and Pavlou, 2014). Moreover, returns are a problem from an ecological perspective (Asdecker et al., 2016; Asdecker and Karl, 2022) in a sector that generally has an impact on the environment (Hill and Lee, 2012). The return of a product and especially the production of the products lead to negative environmental impacts (Sandin et al., 2019). As a result, returns, and in particular returns in the low-priced garments sector, lead to the additional production of apparel, in turn having a negative impact on the environment. However, despite their environmental impact, returns cannot be banned in principle by retail policies to further reduce environmental pollution, as return options are an important driver of consumer confidence and loyalty towards retailers and they represent elements of consumer protection (Yang et al., 2017b). Furthermore, apparel is often referred to as an experience product (Ha and Stoel, 2012). However, touching and trying on garments is excluded when buying online and there are no standardized garment sizes (Pandey and Chawla, 2018) with companies using different measurement systems (Bozzi et al., 2022). Thus, uncertainties about product fit arise, representing a 70 percent share of the reasons for returns in the fashion industry (Lee and Xu, 2020).

For a preventive return management and to reduce orderings of wrong sizes or multiple sizes for selection, many retailers provide the service of offering decision aids. The various aids include, for example, size tables, detailed information on measurements, product illustrations (for example, with photo models), or integrated tools for size advice, representing different innovative approaches (Abdulla and Borar, 2017; Deges, 2017; Nestler et al., 2021). Companies like Zalando for example offer a 3D avatar based on height, weight, and gender (Zalando, 2023) or ASOS provides its fit assistant (ASOS, 2024). Beside of reducing the number of returns, decision aids can further improve consumers' online retailing experience by supporting and guiding the size selection process. While some of these aids try to reduce the amount of articles shown to the customers (Häubl and Trifts, 2000), other decision aids try to support the selection of the right product size (Deges, 2017). Virtual fitting rooms, for example, aim at helping customers deciding for the right product size and can reduce the number of product returns in the post-purchase phase (Yang and Xiong, 2019). Nevertheless, the success of supporting tools depends on how they are used and accepted by customers (Kalpoe, 2020) and how well they support better customer decisions associated with fewer returns.

Many of the decision aids used by retailers require the disclosure of personal consumer data like weight, height or even pictures (ASOS, 2024; Zalando, 2023). However, consumers weigh their risks of disclosing these information with the benefits of getting the best size recommendation (Culnan and Armstrong, 1999). Accordingly, we refer to privacy calculus theory of Laufer and Wolfe (1977) that was also applied to digital contexts in previous research (Dinev and Hart, 2006b). Moreover, we investigate the linkage of privacy concerns and the disclosure of data (Martin and Murphy, 2017) as well as cognitive complexity theory (Kieras and Polson, 1985). Most decision aids represent innovative approaches for consumers, that differ from their known status quo and thus might be rejected due to existing habits and values (Kleijnen et al., 2009). We therefore also regard innovation resistance theory (IRT) (Ram and Sheth, 1989) as important to study the effects of decision aids. While most of the literature has focused on the adoption of innovation like Davis (1989) or Venkatesh et al. (2003), IRT focuses on the rejection of new approaches (Ram and Sheth, 1989).

Summing up our practical and theoretical considerations, we seek to analyze the following research question: How is the intention to use decision aids for online size selection influenced a) by the degree of disclosed data, and b) by their innovativeness? We contribute to research in considering (1) the influence of the degree of the disclosed data on the intention to use them and to disclose data (study I), and (2) the influence of the innovativeness of decision aids for size selection in an online context on the intention to use them and the intention to further transact on the online shop (study II).

3.5.2 Study I

3.5.2.1 Theoretical Background

Study I considers the relevance of the degree of data disclosure in the context of online size selection and investigates consumers' intention to use the decision aid and to disclose their personal information. The conceptual framework for Study I builds on social exchange and privacy calculus theory and is illustrated in Fig. 3.10.

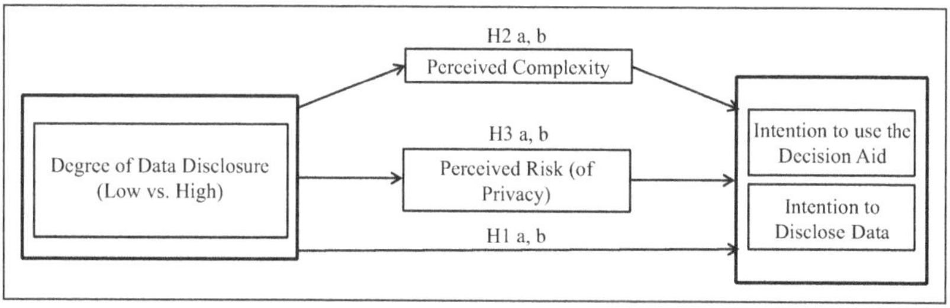

Fig. 3.10 Research Model Study I

Social exchange theory, which can be traced back to several authors from the 1960s and 1970s such as Homans (1958), considers the weighing up of benefits and risks. According to this concept, social behavior involves a continual exchange of resources, whether physical or non-physical (Emerson, 1976; Homans, 1958). A study by Foa and Foa (1974) identified different types of resources being exchanged by participants. One of these types of resources are information (Foa and Foa, 1974). In line with social exchange theory, but in particular investigating the transfer of personal information, privacy calculus theory was established by Laufer and Wolfe (1977). This theory states that consumers weigh the added value of disclosure of personal data against the effort and costs involved, trying to protect their privacy (Engels and Grunewald, 2017). This concept has been considered in the digital world, addressing the use of customer data. Culnan and Armstrong (1999) for example, showed that organizations have to balance the advantages for consumers of sharing their personal information with the privacy concerns arising from. Dinev and Hart (2006b) adapted privacy calculus theory to data disclosure on the internet and describe how modern corporate systems are designed for making the most effective use of customer data, but at the same time pose a certain risk to consumers (Dinev and Hart, 2006b). Consumers thus weigh the benefits of sharing their personal information and data with online shops and decision aids with the risks that this exchange entails. Especially sharing a higher degree of data might be rejected due to the missing equivalent exchange.

Another theoretical approach that considers a similar discussion is the privacy paradox, a term introduced by Norberg et al. (2007). Studies by Brown (2001) or Spiekermann et al. (2001), among others, had already identified this phenomenon. According to these studies, consumers were reluctant and cautious about disclosing private data when shopping online. However, they were willing to disclose their data in return for a service (Brown, 2001) and behaved differently from their previously mentioned attitude (Spiekermann et al., 2001). When disclosing personal information customers become vulnerable and as a result may try to hide these information (Nguyen and Khoa, 2019). Privacy paradox was also observed in the context of online personalization. Awad and Krishnan (2006), for example, showed that consumers that are less willed to be profiled online, value information transparency higher. Anyhow, research shows that customers are interested in using personalized applications and their privacy concerns do not hinder them (Sutanto et al., 2013).

When considering the disclosure of data, uncertainties and risks about privacy thus are relevant both when using technological applications (Kang and Jung, 2021) and when using digital platforms (Liu et al., 2022; Miyazaki and Fernandez, 2001). As a result, most internet users tend to try not to disclose their private data but hide it protecting their privacy (Ardani and Juliadi, 2019), also being shown in the context of e-commerce (Anic et al., 2019) and social electronic commerce (Sun et al., 2019). Hereby, concerns about privacy often are linked to consumers' willingness of disclosing personal data and information (Martin and Murphy, 2017). More specifically related to the context of online decision aids, research regarding the acceptance and success of virtual fitting rooms also identified privacy issues as a main driver of customers' intention to use these fitting rooms (Huang and Qin, 2011; Ivanov et al., 2022). Consumers also show an increased awareness of privacy issues in the context of online decision aids and size recommendations (Miell et al., 2018). Literature, however, has also shown different results in this context. For example, a study by Kim et al. (2019) identified participants to ignore their perceived privacy risks when being offered a better personalized service. Based on the previous theoretical background, we postulate following hypotheses:

H1a: Customers' intention to use the decision aid is negatively influenced by the degree of data disclosure.

H1b: Customers' intention to disclose personal data is negatively influenced by the degree of data disclosure.

Considering the adoption and use of technologies or digital applications, the Technology Acceptance Model (TAM) of Davis (1989) and the Unified Theory of Acceptance and Use of Technology (UTAUT) by Venkatesh et al. (2003) are two of the most applied theoretical approaches. Referring to these theories, perceived complexity plays a decisive role and can be derived from various of the variables considered in the models. While perceived complexity focuses on how difficult an application is to understand or use (Thompson et al., 1991), the TAM considers how easy a technology is to use (Davis, 1989). As the intention to use a technology increases through perceived ease of use (Davis, 1989), it is likely that the intention to use decreases with increasing complexity. In addition, Venkatesh et al. (2003) consider effort expectancy in their UTAUT and derive this determinant from perceived complexity of Thompson et al. (1991). Accordingly, we expect a negative influence of perceived complexity on the intention to use a new application. Recent research has shown that the complexity of recommendation agents has a negative impact on customers' intention to use them (Ghasemaghaei et al., 2019). If the complexity of information when disclosing additional data exceeds the information processing capacity of customers, this in turn can lead to negative consequences (Dabholkar, 2006). Todd and Benbasat (2000) found that customers mainly use decision aids in an online shopping context to reduce the effort required to make a decision. This goes in line with cognitive complexity theory, according to which users reject technologies if they require too much additional cognitive input and are therefore perceived as complex (Kieras and Polson, 1985). Referring to that, a higher degree of data disclosure required for using online decision aids is expected to increase customers' perceived complexity. We expect perceived complexity to mediate the effect of the degree of data disclosure on customers' intention to use the decision aid and to disclose their private data. Drawing on the theoretical considerations, we propose the following hypotheses:

H2a: The effect of the degree of data disclosure on customers' intention to use the decision aid is mediated by perceived complexity.

H2b: The effect of the degree of data disclosure on customers' intention to disclose personal data is mediated by perceived complexity.

Following the approach of privacy calculus theory, higher levels of data required for using decision aids, require higher benefits for consumers in exchange (Dinev and Hart, 2006b). Consumers trade off the benefits of sharing personal information with the privacy related risks, influencing their intention of disclosing personal information in a digital context (Wang et al., 2016). Furthermore, Li et al. (2016) considered this risk-benefit analysis by referring to privacy calculus. The authors showed an influence of information sensitivity on perceived privacy risk. Hereby, information sensitivity is characterized by the type of information and the level of information required, similar to the degree of data disclosure investigated in our study (Li et al., 2016). This confirmed results by Dinev and Hart (2006a), who showed an increasing level of information exchange leading to greater associated privacy concerns. Furthermore, the authors measured a negative relationship between privacy concerns and consumers intention of using the service (Dinev and Hart, 2006a). Based on these studies, higher levels of required data increase consumers perceived privacy risk, negatively influencing consumers' intention to use the decision aids and to disclose their private data. Therefore, we postulate a mediating effect of perceived privacy risk in this context.

H3a: The effect of the degree of data disclosure on customers' intention to use the decision aid is mediated by perceived risk of privacy.

H3b: The effect of the degree of data disclosure on customers' intention to disclose personal data is mediated by perceived risk of privacy.

3.5.2.2 Empirical Study: Method and Procedure

To analyze the relevance of the degree of data disclosure when using online decision aids for size selection, an experiment was conducted, using a between-subject design with two scenarios. All subjects were told to remember their last visit of an apparel online store and a fictitious online store was presented. By differentiating in female and male subjects, a gender-specific, however rather neutral, shirt was depicted, of which the customers were requested to select the size. The scenario is illustrated in Fig. 3.11.

Fig. 3.11 Illustration of the Online Shop for both Scenarios for Men

As a support for the size selection of the shirt, each subject was presented one of two decision-making aids via written and graphical presentation. The two scenarios differentiated in the way that customers had to upload a picture of themselves additionally to data that had to be entered in both scenarios. Referring to Ivanov et al. (2022), the upload of pictures represents an increased degree

Fig. 3.12 Illustration of Scenario 1

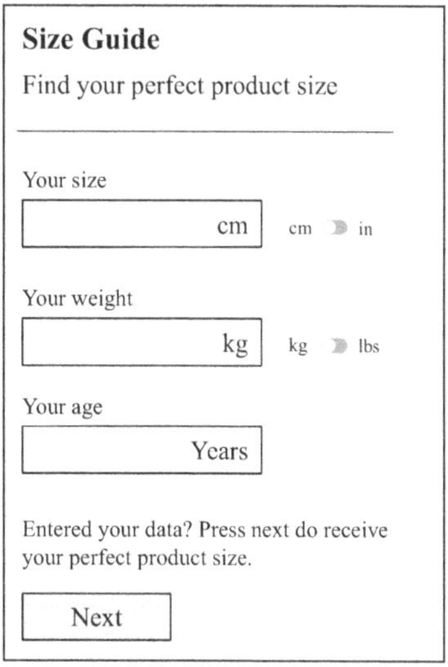

of data disclosure for consumers and thus is suited for our manipulation. The decision aids were only presented in a purely informative way and participants did not have to input any data about their body size or a picture themselves. In the first scenario, participants had to enter their body weight, height, and age. Based on this, their appropriate size was suggested by the decision aid. In the second scenario, participants also had to enter their body weight, height, and age. Additionally, they had to upload a picture of them for getting a more precise result of the appropriate size. Participants were shown an example of how they should place their bodies on this picture. This example was differentiated in male and female. The models used for these examples had their faces blurred. Both scenarios are depicted in Fig. 3.12 and 3.13, using the version for a male consumer as an example.

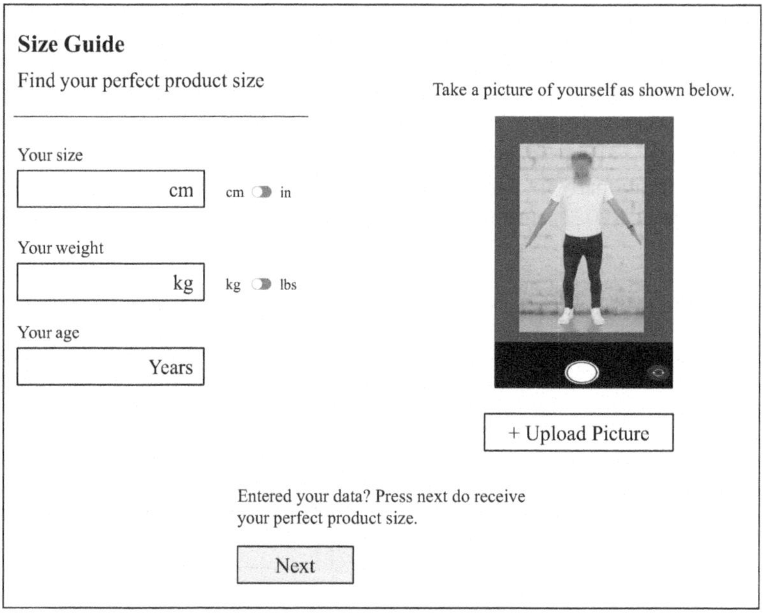

Fig. 3.13 Illustration of Scenario 2

Regarding the dataset, it was ensured that the participants were equally assigned to the two scenarios (low data disclosure: 72; high data disclosure: 66). To limit the number of scenarios, we could only include female and male subjects in our study and could not include other genders. A total of 138 participants (women: 63 %) with an average age of 27.4 years (SD = 9.76) were included in the final study. To test the hypotheses, established constructs were adapted to this study, all measured on a seven-point-Likert-scale ranging from "not apply at all" (1) to "fully apply" (7). These constructs are illustrated in Table 3.17. Reliability was measured by calculating Cronbach's Alpha (α) meeting the required value (α ≥ .7) for all constructs (Hair et al., 2019).

Table 3.17 Constructs and Items used in Study 1

Construct (source)	Item	α
Intention to Use (Wang and Benbasat, 2009)	Assuming I had access to the system, I intend to use the decision aid.	.988
	Assuming I had access to the system, I predict that I will use the decision aid.	
	Assuming I had access to the system, I plan to use the decision aid.	
Intention to Disclose Data (Wakefield, 2013)	Would you share the required data with the system?	.950
	Probably	
	Presumably	
	Possibly	
	Willingly	
Perceived Privacy Risk (Kang and Jung, 2021)	In general, it is risky to enter my personal data in the size advisor.	.939
	Entering my personal data in the size advisor is associated with a high risk.	
	Entering my personal data in the size advisor is associated with too much uncertainty.	
	Entering my personal data in the size advisor could lead to unexpected problems.	
Perceived Complexity (Thompson et al., 1991)	Using the size advisor takes too much time.	.878
	Working with the size advisor is so complicated that it is difficult to understand what is going on.	

(continued)

Table 3.17 (continued)

Construct (source)	Item	α
	Using the size advisor requires too much time for mechanical operations (for example, data entry).	
	It takes too long to learn how to use the sizes advisor to make it worth the effort.	
Degree of Data Disclosure (Fan et al., 2014)	The degree of information shared with the decision aid is high.	.728
	The degree of information provided to the decision aid is enough.	
	The electronic degree of information sharing is high.	

Participants were also asked questions to check whether they had really read and understood the scenarios. 23 participants with not matching answers were excluded from the final study. No distortions were found for technology affinity or experience with the use of online stores between the scenarios. The gender of the survey participants also has no influencing effect. ANOVA tests were calculated, confirming a successful manipulation of the level of data disclosure within the experiment (Data Disclosure: $M_{\text{low data disclosure}} = 4.73$ (1.31), $M_{\text{high data disclosure}} = 5.60$ (1.08), $F(1.136) = 18.083$, $p < .001$, $\eta^2 = 0.12$). Results for the low degree of data disclosure already show a value of 4.73, indicating that consumers perceive the entry of even a small amount of private data as a higher level of data entry. Despite the high value for the low degree of data disclosure, significant differences between the scenarios are measured.

3.5.2.3 Findings

Data was analyzed by using the software SPSS. First, ANOVAs were calculated, showing consumers' intention to use the decision aid (Intention to Use: $M_{\text{low data disclsoure}} = 4.25$ (1.55), $M_{\text{high data disclsoure}} = 3.28$ (1.91), $F(1.136) = 10.656$, $p < .001$, $\eta 2 = 0.073$) and consumers' intention to disclose data (Intention to Disclose Data: $M_{\text{low data disclsoure}} = 3.73$ (0.91), $M_{\text{high data disclsoure}} = 2.91$ (1.23), $F(1.136) = 20.199$, $p < .001$, $\eta 2 = 0.129$) to be significantly higher with a lower degree of data disclosure required. Furthermore, results of a linear regression confirmed that the degree of data disclosure when using the decision aid has a negative effect on the intention to use ($R^2 = .073$, $ß = -.270$, $p < .001$) and on the intention to disclose data ($R^2 = .129$, $ß = -.360$, $p < .001$). Results confirm hypotheses 1 a and b.

To test the postulated mediating influence of perceived complexity and perceived privacy risk on the effects of the experimental factor on the dependent variables, PROCESS (model 4) was used, as suggested by Zhao et al. (2010). Following the approach of Baron and Kenny (1986) and their first of four steps of mediation, significant values for the direct paths of the experimental factor on the dependent variables (Intention to Use: ß = −.963, t(136) = −3.26, p = .001; Intention to Disclose Data: ß = −.824, t(136) = −4.49, p < .001) could be measured. Further results of the four mediation analyses are presented in Table 3.18.

Table 3.18 Mediation Results Study 1

IV	DV	ß	T	F	R	R2
Degree of Data Disclosure	Perceived Complexity	.81***	3.82	14.619***	.312	.097
Degree of Data Disclosure	Intention to Use	−.628*	−2.11	12.333***	.393	.155
Perceived Complexity		−.41***	−3.61			
Degree of Data Disclosure	Intention to Disclose Data	−.550**	−3.09	23.974***	.512	.262
Perceived Complexity		−.337***	−4.93			
Degree of Data Disclosure	Perceived Privacy Risk	1.12***	4.59	21.038***	.366	.134
Degree of Data Disclosure	Intention to Use	−.38	−1.32	22.004***	.496	.246
Perceived Privacy Risk		−.52***	−5.57			
Degree of Data Disclosure	Intention to Disclose Data	−.35*	−2.16	44.998***	.632	.400
Perceived Privacy Risk		−.42***	−7.80			

DV = dependent variable, IV = independent variable
Sig.:*p < .05,**p < .01,***p < .001

As shown in Table 3.18, significant paths for the degree of data disclosure on perceived complexity and perceived privacy risk as well as significant paths from perceived complexity and perceived privacy risk on intention to use the decision aid and on intention to disclose data could be measured, confirming step two and three of Baron and Kenny (1986). Split results emerge for step four. Regarding the mediating effect of perceived complexity, both the indirect and direct paths of data disclosure on the intention to use are significant, thus indicating partial mediation (relative total effect: $ß = -.96$, $p = .001$; relative direct effect: $ß = -.63$, $p = .04$). The relative indirect effect from the experimental factor on perceived complexity on intention to use the decision aid shows a significant mediation (relative indirect effect = −.33, SE = .14, 95% CI[−.63, −.10]). Results partly confirm hypothesis 2a. Considering the mediating effect of perceived complexity, both the indirect and direct paths of data disclosure on the intention to disclose data are significant, also indicating partial mediation (relative total effect: $ß = -.82$, $p < .001$; relative direct effect: $ß = -.55$, $p = .003$). The relative indirect effect from the experimental factor on perceived complexity on intention to disclose data shows a significant mediation (relative indirect effect = −.27, SE = .10, 95% CI[−.50, −.10]). Results partly confirm hypothesis 2b. Regarding the mediating effect of perceived privacy risk, only the indirect path of data disclosure on the intention to use remains significant, thus indicating full mediation (relative total effect: $ß = -.96$, $p = .001$; relative direct effect: $ß = -.38$, $p = .19$). The relative indirect effect from the experimental factor on perceived privacy risk on intention to use the decision aid shows a significant mediation (relative indirect effect = −.58, SE = .17, 95% CI[−.97, −.28]). Results confirm hypothesis 3a. Finally, considering the mediating effect of perceived privacy risk, both the indirect and direct paths of data disclosure on the intention to disclose data are significant, thus indicating partial mediation (relative total effect: $ß = -.82$, $p < .001$; relative direct effect: $ß = -.35$, $p = .03$). The relative indirect effect from the experimental factor on perceived complexity on intention to use the decision aid shows a significant mediation (relative indirect effect = −.47, SE = .13, 95% CI[−.75, −.24]). Results partly confirm hypothesis 3b.

3.5.2.4 Discussion

Results show that higher levels of data required for online decision aids lead to lower usage intentions and lower intentions to disclose data. These results go in line with research of Dinev and Hart (2006b), according to which privacy concerns have increased in times of online shopping. Referring to privacy calculus theory, consumers do not value the benefits of the decision aid high enough to disclose a higher degree of data (Culnan and Armstrong, 1999). In addition, we could show that a higher level of data disclosure in decision support leads to the perception of the decision aid as being more complex and having a negative impact on the intention to use the support. Consumers seem to have to exert more effort (Engels and Grunewald, 2017) and invest more time (Wang and Benbasat, 2009) resulting in higher perceived complexity negatively mediating the effect of the degree of data disclosure and usage intention. Results are in line with Tornatzky and Klein (1982) also identifying a negative effect of perceived complexity on new applications.

Furthermore, mediation by perceived privacy risk is consistent with previous research about internet users trying to hide their private data on the internet (Ardani and Juliadi, 2019) and about concerns that privacy is negatively influenced by data disclosure (Wang and Liu, 2014). The awareness of privacy issues was already investigated in the context of online decision aids and stated similar results (Miell et al., 2018). The influence of information sensitivity on perceived privacy risk (Dinev and Hart, 2006a; Li et al., 2016) was supported by our results, showing an increase of perceived privacy risk.

3.5.3 Study II

3.5.3.1 Theoretical Background

Having figured out that the degree of data disclosure when using decision aids negatively affects consumers' usage and disclosure intentions in Study I, the focus in Study II shifts to examining how innovativeness of these decision aids influences consumer behavior. Specifically, the influence of innovativeness on consumers intention to use the decision aid and on the intention to transact in the online shop in general will be investigated. The research model for Study II is illustrated in Fig. 3.14.

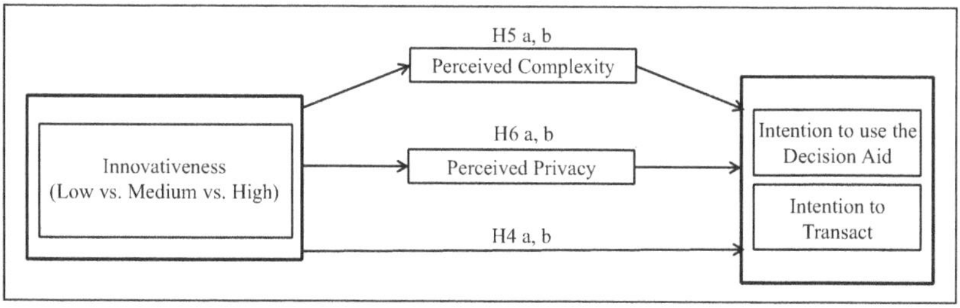

Fig. 3.14 Research Model for Study II

Literature investigating the adoption of new technological approaches has mainly focused on the acceptance of innovations like the TAM (Davis, 1989) or the UTAUT (Venkatesh et al., 2003). However, compared to the acceptance of new technological approaches, literature has rarely focused on their rejection (Claudy et al., 2015). Individuals generally show resistance to change and exhibit a reluctance to embrace change (Oreg, 2003). Based on this, new technologies have slow adoption rates and consumers are not willed to change their established behaviors (Claudy et al., 2015) and even reject innovations (Ram and Sheth, 1989). Especially resistance by customers has been identified as one of the main reasons why innovations could not establish in the long term (Kalpoe, 2020). Ram and Sheth (1989) have established IRT, stating that individuals reject the adoption of innovative approaches and prefer maintaining the give status quo. Consumers know that existing solutions are successful and therefore want to preserve the current standard (Gourville, 2006). Kaur et al. (2020) list various aspects of why customers might resist innovations. Beside the change from the current status quo, customers need to learn how to use an innovative technology or innovations may not conform to customers' traditions or norms. Innovativeness of technologies negatively affects customers' intention to use them (Ram and Sheth, 1989). Chang and Tung (2008) have shown that there is a strong relationship between innovativeness of a technology and behavioral intention to use this technology. Innovative approaches are often accompanied by customers' uncertainties about the adoption of the innovation (Rogers, 1995), representing a main reason for rejection (Herzenstein et al., 2007). If innovations are consistent with customers' values and habits, they are more likely to be adopted (Tornatzky and Klein, 1982). The greater the changes of new technological approaches from known applications, the greater resistance to them will be (Laukkanen et al., 2007). While higher degrees of innovativeness usually differ more from the current habits and values and are more often rejected by consumers (Kleijnen et al., 2009), we expect more innovative decision aids to be less adopted by consumers. Following hypotheses are formulated:

H4a: Customers' intention to use the decision aid is negatively influenced by the innovativeness of the decision aid.

H4b: Customers' intention to transact is negatively influenced by the innovativeness of the decision aid.

An early meta-analysis by Tornatzky and Klein (1982) has shown that the complexity of an innovation is one of the characteristics influencing its adoption the most. When an innovation is perceived to be more complex, the rate of adoption decreases (Thompson et al., 1991; Tornatzky and Klein, 1982). According to cognitive complexity theory (Kieras and Polson, 1985), users reject new technologies if they require too much additional cognitive input and are therefore perceived as complex. Similar to cognitive complexity theory, Todd and Benbasat (2000) found that customers mainly use decision aids in an online shopping context to reduce the effort required to make a decision. Online decision aids can help customers to mitigate through available product related information (Virdi et al., 2020). But if the amount and complexity of information exceed the information processing capacity of customers, this in turn can lead to negative consequences (Dabholkar, 2006). Thus, we expect more innovative decision aids to increase perceived complexity and thus negatively affect consumers' user intention and intention to transact.

H5a: The effect of the innovativeness on customers' intention to use the decision aid is mediated by perceived complexity.

H5b: The effect of the innovativeness on customers' intention to transact is mediated by perceived complexity.

Referring to the results of Study I of this paper, data disclosure and consumers' concerns about privacy have been shown highly relevant for consumers' intention of using decision aids. According to privacy calculus theory, consumers trade of the benefits and risks of sharing personal information when using a technological approach (Dinev and Hart, 2006b; Wang and Benbasat, 2016). Following IRT (Ram and Sheth, 1989), innovative approaches are linked to functional risk barriers, arising due to uncertainties as well as psychological barriers of risks associated with these innovations (Kaur et al., 2020; Kleijnen et al., 2009). Based on this and the results of Study I, the mediating effect of perceived privacy will be investigated. Following hypotheses are postulated:

H6a: The effect of the innovativeness on customers' intention to use the decision aid is mediated by perceived privacy.

H6b: The effect of the innovativeness on customers' intention to transact is mediated by perceived privacy.

3.5.3.2 Empirical Study: Method and Procedure

To analyze the influence of innovativeness of the decision aid, we conducted a second online experiment, using a between-subject design with three scenarios. All subjects were told to remember their last visit of an apparel online store. In addition, a fictitious online store was presented. By differentiating in female and male subjects, a gender-specific, however rather neutral, shirt was depicted, of which the customers were requested to select the size (Fig. 3.15).

Fig. 3.15 Illustration of the Online Shop for the three Scenarios for Men

To investigate the impact of innovativeness on decision-making, each subject was randomly assigned one of three increasingly innovative decision-making aids on the following page. The written scenarios were enhanced with graphical illustrations to provide a clearer understanding. In the first, more traditional scenario, subjects were given basic information: a model abstractly depicted wearing the t-shirt and detailed circumference measurements to help guide their decision (Fig. 3.16). This served as the first condition, representing a low-innovative decision aid. The second scenario introduced a moderately innovative aid. Respondents were required to input data such as age and body proportions into the decision aid, which then calculated the appropriate size (Fig. 3.17).

This scenario incorporated basic personalization features, marking a step up in innovativeness. The third scenario presented a highly innovative decision aid, where subjects had to enter similar personal data but also upload a short video of a body rotation (Fig. 3.18). This advanced feature utilized video analysis to provide a more precise size recommendation, representing the highest level of innovation among the aids. Participants did not actually input any data or upload videos themselves. Instead, they were informed about the procedures through written descriptions and illustrations, allowing us to manipulate the perceived innovativeness of each decision-making aid without requiring active participation.

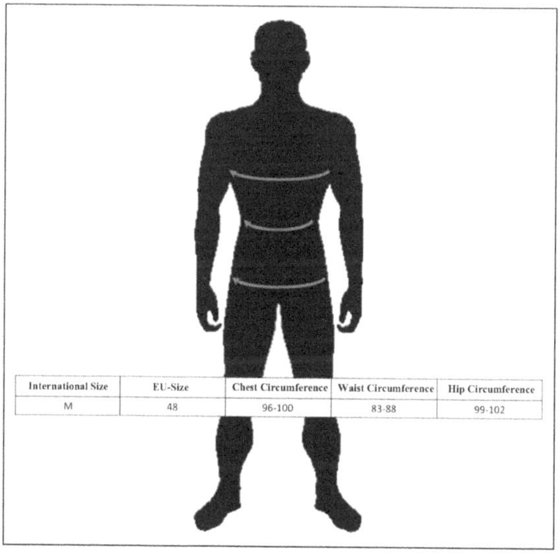

Fig. 3.16 Illustration of Scenario 1

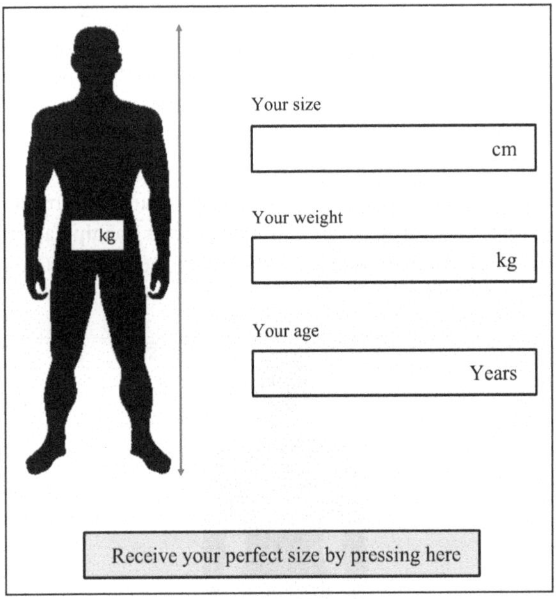

Fig. 3.17 Illustration of Scenarnio 2

It was assumed that the three gradations of the scenarios systematically manipulated the degree of innovation of the decision aids, with scenario one representing the lowest innovation and scenario three the highest. Decision aids have been provided for many years based on circumferences, while the use of automated recommendations based on entered data is a new process that differs significantly from these existing processes and therefore represents an innovative solution. The addition of the body scan makes this example a very innovative solution compared to technologies currently used. The manipulation was tested in a pre-test.

Regarding the dataset, it was ensured that the participants were equally divided between the different scenarios (low innovativeness: 43; medium innovativeness: 45; high innovativeness: 43). A total of 131 participants (women: 71 %) with an average age of 25.7 years (SD = 6.73) participated in the study.

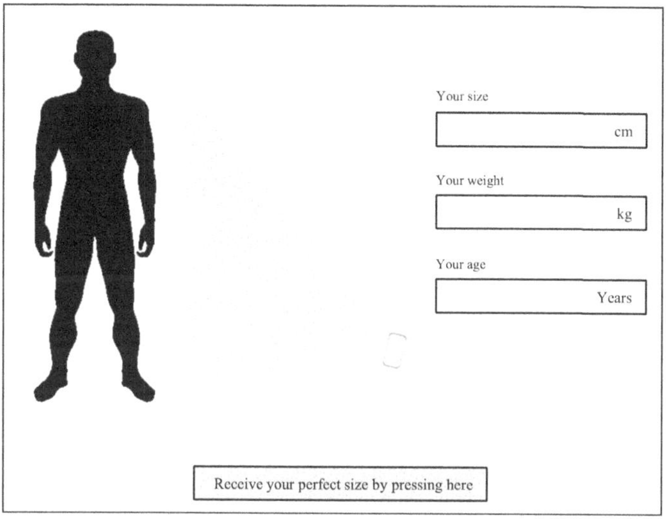

Fig. 3.18 Illustration of Scenario 3

To test the hypotheses, established constructs were adapted to this study, all measured on a seven-point-Likert-scale ranging from "not apply at all" (1) to "fully apply" (7) and calculated by mean-aggregation. These constructs are illustrated in Table 3.19. Reliability was measured by calculating Cronbach's Alpha (α). The required value (α ≥ .7) was achieved for all measurements (Hair et al., 2019).

Table 3.19 Constructs and Items used in Study 2

Construct (source)	Item	α
Perceived Complexity (Venkatesh et al., 2003)	Using the size advisor takes too much time.	.902
	Working with the size advisor is so complicated that it is difficult to understand what is going on.	
	Using the size advisor requires too much time for mechanical operations (for example, data entry).	
	It takes too long to learn how to use the size advisor to make it worth the effort.	

(continued)

Table 3.19 (continued)

Construct (source)	Item	α
Perceived Privacy (Blut, 2016)	I trust the website to keep my information safe.	.956
	I trust the website administrators to not misuse my information.	
	It protects information about my web-shopping behavior based on using the decision aid.	
Intention to Use (Wang and Benbasat, 2009)	Assuming I had access to the system, I intend to use the decision aid.	.971
	Assuming I had access to the system, I predict that I will use the decision aid.	
	Assuming I had access to the system, I plan to use the decision aid.	
Intention to Transact (Pavlou, 2003)	Given the chance, I intend to use this online shop.	.974
	Given the chance, I predict that I should use this online shop in the future.	
	It is likely that I will transact with this online shop in the near future.	
Innovativeness (Loiacono et al., 2002)	The decision aid is innovative.	.926
	The decision aid is creative.	
	The conception of the decision aid is innovative.	

No distortions were found for technology affinity or experience with the use of online stores between the scenarios. Since the scenarios were customized to the gender of the survey participants, an influencing factor of gender was also checked for. However, no differences were measured and thus, the scenarios analyzed together. ANOVA tests were calculated, confirming a successful manipulation within the experiment (Innovativeness: $M_{\text{low innovativeness}} = 2.84$ (1.34), $M_{\text{medium innovativeness}} = 4.03$ (1.72), $M_{\text{high innovativeness}} = 4.99$ (1.63), $F(2.128) = 20.193$, $p < .001$, $\eta^2 = 0.24$). Tukey post-hoc test also confirmed that innovativeness differs significantly across all scenarios.

3.5.3.3 Findings

The experimental factor innovativeness is represented by a variable with three conditions (low vs. medium vs. high innovativeness). Accordingly, ANOVA tests were calculated to determine the varying usage intentions (Intention to use: $M_{low\ innovativeness}$ = 4.72 (1.65), $M_{medium\ innovativeness}$ = 3.64 (1.79), $M_{high\ innovativeness}$ = 3.21 (1.87), F(2.128) = 8.306, p < .001, η2 = 0.115) and intentions to transact (Intention to transact: $M_{low\ innovativeness}$ = 3.88 (1.62), $M_{medium\ innovativeness}$ = 3.41 (1.47), $M_{high\ innovativeness}$ = 3.64 (1.63), F(2.128) = .976, p = .38, η2 = 0.015) of the three scenarios. Variance homogeneity and an equal sample size were ensured, allowing for a Tukey post-hoc test to be carried out, examining significant differences of consumers' intention to use the decision aid between the three scenarios. Results indicate that scenario 1 differed significantly from scenario 2 and from scenario 3. No significant differences are observed between scenarios 2 and 3 in terms of usage intention. There is a trend showing that the usage intention decreased as innovativeness increased. In general, the intention to utilize the decision aid diminishes with increasing innovativeness. Results of a linear regression confirm that the innovativeness of the decision aid has a negative effect on the intention to use (R2 = .108, ß = −.328, p < .001). Accordingly, H4a can be confirmed. Results of the ANOVA and Tukey post-hoc test considering consumers' intention to transact, no significant differences between the three scenarios could be measured. This is further supported by a linear regression (R2 = .004, ß = −.06, p = .495). H4b must be rejected. The mediating effects of perceived complexity and privacy on the innovativeness of the decision aid on consumers' intention to transact are not investigated.

To test the postulated mediating influence of perceived complexity and privacy on the effect of innovativeness on consumers' intention to use the decision aid, PROCESS (model 4) was used, as suggested by Zhao et al. (2010). Following the approach of Baron and Kenny (1986) and their first of four steps of mediation, significant values for the direct paths of the experimental factor on the dependent variable (X1: ß = −1.08, t(128) = −2.86, p = .005; X2: ß = −1.51, t(128) = −3.95, p < .001) could be identified. Since the experimental factor represents a multi-categorical variable in PROCESS, path X1 represents the ratio medium innovativeness vs. low innovativeness and path X2 represents the ratio high innovativeness vs. low innovativeness. Further results of the two mediation analyses are presented in Table 3.20.

Table 3.20 Mediation Results Study 2

IV	DV	ß	T	F	R	R2
X1	Perceived Complexity	1.05***	3.16	11.666***	.393	.154
X2		1.60***	4.75			
X1	Intention to Use	−.60	−1.67	14.766***	.508	.259
X2		−.78*	−2.04			
Perceived Complexity		−.46***	−4.96			
X1	Perceived Privacy	−.25	−.65	1.608	.157	.025
X2		−.70	−1.77			
X1	Intention to Use	−1.00**	−2.81	12.713***	.481	.231
X2		−1.27***	−3.50			
Perceived Privacy		.35***	4.38			

DV = dependent variable, IV = independent variable
Sig.:*p < .05,**p < .01,***p < .001

As shown in Table 3.20, significant paths for X1 and X2 on perceived complexity and from perceived complexity on intention to use the decision aid could be measured, confirming step two and three of Baron and Kenny (1986). Split results emerge for step four. While the path of X1 on the dependent variable is no longer significant among the addition of the mediator, thus indicating full mediation, the path of X2 on the dependent variable still shows weak significance, thus indicating partial mediation (relative total effect: X1: ß = −1.08, p = .05, X2: ß = −1.51, p < .001; relative direct effect: X1: ß = −.60, p = .10, X2: ß = −.78, p = .04). The relative indirect effect from the experimental factor on perceived complexity on intention to use the decision aid shows mediation for both group comparisons (X1 indirect = −.48, SE = .19, 95% CI[−.91, −.16]; X2 indirect = −.73, SE = .21, 95% CI[−1.18, −.35]. Results partly confirm hypothesis 5a.

Considering the results of the mediation of perceived privacy, the paths for X1 and X2 on perceived privacy are not significant. Following Baron and Kenny (1986), hypothesis 6a has to be rejected. However, as shown in Table 3.20, perceived privacy significantly influences consumers' intention to use.

3.5.3.4 Discussion

Results of Study II generally support the basic assumptions of the IRT (Ram and Sheth, 1989). Innovativeness lowers consumers' intention to use decision aids.

No significant effect of innovativeness on consumers' intention to transact could be measured. Results indicate that innovative decision aids would discourage consumers to use these aids, but not influence the transaction in the online shop in general. When having greater changes to the known status quo, consumer resistance seems to be higher, going in line with literature (Laukkanen et al., 2007).

Perceived complexity mediates the effect of innovativeness on consumers' intention to use decision aids. According to the literature, the use of more innovative decision aids requires too much cognitive input and is perceived as more complex (Kieras and Polson, 1985), exceeding their information processing capacity (Dabholkar, 2006). Interestingly, no significant mediation for perceived privacy could be measured. The innovativeness of the decision aid seems not to influence consumers' perceived privacy. However, privacy related concerns remain an important determinant for consumers intention to use the decision aid, as a significant effect of perceived privacy on consumers intention to use the decision aid was measured.

3.5.4 General Discussion and Implications

3.5.4.1 Theoretical Discussion

Table 3.21 provides an overview of the acceptance or rejection of the postulated hypotheses. In general, results have shown that the degree of data disclosure and innovativeness both negatively affect consumers' intention to use decision aids. Significant results for perceived complexity and risk could be further measured.

Table 3.21 Summary of the Results

IV	Med.	DV	H	Result
Degree of Data Disclosure		Intention to Use	1a	confirmed
Degree of Data Disclosure		Intention to Disclose	1b	confirmed
Degree of Data Disclosure	Perceived Complexity	Intention to Use	2a	partly confirmed
Degree of Data Disclosure	Perceived Complexity	Intention to Disclose	2b	partly confirmed

(continued)

Table 3.21 (continued)

IV	Med.	DV	H	Result
Degree of Data Disclosure	Perceived Risk	Intention to Use	3a	confirmed
Degree of Data Disclosure	Perceived Risk	Intention to Disclose	3b	partly confirmed
Innovativeness		Intention to Use	4a	confirmed
Innovativeness		Intention to Transact	4b	rejected
Innovativeness	Perceived Complexity	Intention to Use	5a	partly confirmed
Innovativeness	Perceived Complexity	Intention to Transact	5b	rejected
Innovativeness	Perceived Privacy	Intention to Use	6a	rejected
Innovativeness	Perceived Privacy	Intention to Transact	6b	rejected

Various theoretical approaches can be derived from the results of our empirical studies. The findings indicate that an increased level of data disclosure has a harmful impact on consumers' intention to use decision aids and to share their personal information. In accordance with the assumptions of privacy calculus theory (Dinev and Hart, 2006b; Laufer and Wolfe, 1977), the perceived costs to consumers, including uncertainty, effort, and security concerns, outweigh the perceived benefits of disclosing data (Engels and Grunewald, 2017). The benefits of the decision aids are not perceived valuable enough by consumers to justify the disclosure of a higher degree of data. Furthermore, the balance between advantages and risks is not perceived as favorable (Culnan and Armstrong, 1999). Results indicate that the perceived privacy risk associated with entering personal data into the decision aid fully mediates consumers' intention to use the decision aid and partly mediates consumers' intention to disclose their data. This is consistent with findings of Dinev and Hart (2006a), who also demonstrated that increased information exchange is associated with heightened privacy concerns. Moreover, the results align with those of other studies that have underscored the relevance of privacy issues in the context of virtual fitting rooms (Huang and Qin, 2011; Ivanov et al., 2022) and highlighted privacy concerns as a key reason for not disclosing private data (Ardani and Juliadi, 2019). In contrast to the findings of Kim et al. (2019), which indicated that privacy risks are ignored in pursuit of more personalized services, our results do not support this conclusion. Perceived complexity has been identified as a mediator for the effects of the degree of data disclosure on consumers' intention to use the decision aid and their intention to disclose personal data. Our results are consistent with the findings of Venkatesh et al. (2003) and Tornatzky and Klein (1982), which indicate that perceived complexity has a negative impact on the adoption of new technologies and approaches. Moreover, our results align with previous research on complexity of recommendation agents (Ghasemaghaei et al., 2019). According to cognitive complexity theory, the necessary additional input increases perceived complexity (Kieras and Polson, 1985), representing a challenge for the use of decision aids (Todd and Benbasat, 2000).

Literature on the adoption of new technological approaches has mainly focused on the acceptance of these innovations, with less attention paid to their rejection (Claudy et al., 2015). However, a theoretical approach that investigates the rejection of innovation is called IRT (Ram and Sheth, 1989). The results of the studies lend support to IRT, indicating that as decision aids become more innovative, consumers' intention to utilize these aids declines. It is likely that more innovative decision aids will diverge significantly from current consumer habits and values (Kleijnen et al., 2009), and will be accompanied by a certain degree of uncertainty (Rogers, 1995). Interestingly, the intention to conduct a transaction in the online shop is not influenced by levels of innovativeness. Therefore, consumers are reluctant to utilize more innovative decision aids, yet their intention to purchase items from an online store remains unchanged due to this innovativeness. Results are consistent with previous research on the role of complexity in the context of innovations, which have demonstrated a negative influence on user adoption (Thompson et al., 1991; Tornatzky and Klein, 1982). No substantial mediation of perceived privacy could be observed.

3.5.4.2 Practical Implications

Several practical implications can be derived from the results of the studies. It would be prudent for retailers to consider the degree of data that is required for utilization of online decision-aids. Required input of data by consumers should be minimal and the usage should be straightforward and uncomplicated. Moreover, retailers should provide accurate information about the security, use, and retention of disclosed data to address customers' privacy concerns (Culnan and Armstrong, 1999). In accordance with the tenets of privacy calculus theory (Dinev and Hart, 2006b), retailers should strive to achieve a balance between the advantages of a personalized service when using the decision aid and the potential risks associated with the disclosure of personal data to the aid (Culnan and Armstrong,

1999). In cases where a greater degree of data disclosure is necessary, it is essential to clearly highlight the associated benefits to achieve a balanced outcome. Moreover, retailers may opt to incrementally enhance their decision aid portfolio, thereby avoiding significant deviations from the current status quo. It is important for retailers to be prepared for the possibility of consumers initially rejecting new, innovative decision aids. The launch of new products should be accompanied by comprehensive explanations of the decision-aiding tools, instructions for their use, and information about the data they generate. This will help to reduce consumers' uncertainty and increase the rate of adoption. In general, retailers should consider whether they can offer consumers a clear advantage when providing decision-aiding tools in their online shops or whether they should focus on other benefits for consumers. According to the literature, retailers should offer additional explanations about the size recommendations to convince consumers to use these decision aids (Tan et al., 2012). In the context of online shopping, it is crucial for retailers to consider the complexity of decision aids for consumers. The aids should not require excessive effort from users and should not be too complex. Additionally, retailers should not force consumers to use the decision aids, but rather present them as an option for usage.

3.5.4.3 Limitations and Future Research

The two studies have several limitations that should be taken into account when considering the results. Online experiments were conducted and should be verified in a real experiment. In addition, the average age of participants in both studies is relatively low. This might influence the perception of technology usage. Many of the participants never have used any of the decision aids mentioned in the studies, making it difficult for them to understand their functions and usage options. For a better assessment of participants' understanding of the functions of the decision aids, appropriate attention checks were used. The key findings should be further explored for investigating what type of decision aid consumers would be most likely to use. It would also be particularly essential to conduct a longitudinal analysis. The decision aids lose their degree of innovation over time because customers become more and more used to them. Effects over time would be particularly relevant to investigate. Previous research already identified differences between cognitively younger and older adults (Ghasemaghaei et al., 2019) as well as the level of media richness in e-commerce (Maity and Dass, 2014), which could also play a role for the use of decision aids. Visual cues are an important aspect in online retailing and should be implemented in future experiments using visualizations of online shops (Roberts and Grassi, 2021). In addition, privacy calculus theory implies that customers are cautious to share their personal information on the internet, for example during online shopping. However, the actual usage seems ambivalent with steadily increasing e-commerce orders (Dinev and Hart, 2006b) and should be investigated in further research in the context of online decision aids. Referring to privacy calculus theory (Dinev and Hart, 2006b), the conflict between benefits and risks should be considered in more detail. When consumers are required to upload images of themselves in order to obtain recommendations regarding the optimal size, self-esteem might be a significant influencing factor (Yang and Xiong, 2019) and can be decomposed into performance, social, and appearance (Heatherton and Polivy, 1991), being relevant for future research.

3.6 Essay 6—Sustainability Assessment of Last-Mile Electrification: A Qualitative Study in Germany

This study was published as: Selter, J. L.; Schmitz, J. & Schramm-Klein, H., 2024. Sustainability assessment of last-mile electrification: A qualitative study in Germany. Transportation Research Part D: Transport and Environment 126, 104019.

3.6.1 Introduction

Climate change is one of the greatest challenges of our time, and a commitment to decarbonization to preserve the environment was signed by 195 countries in the Paris Climate Agreement (Glanemann et al., 2020). CO_2 emissions have a major impact on climate change (Quak et al., 2016), and the transportation sector produces one-fifth of global CO_2 emissions (Ritchie, 2020). The last mile in particular is a major source of CO_2 pollution but has steadily grown in the past few years because of increasing e-commerce and home deliveries (Kaplan et al., 2016; Wolff and Madlener, 2019). With over 60 % of the global population projected to live in urban areas, emissions' impact will be most felt in these areas (United Nations, Department of Economic and Social Affairs Population Division, 2019). Urban transportation disrupts cities with issues like noise, air pollution, and traffic congestions (Juan et al., 2016; Quak et al., 2016). Moreover, bans on certain vehicle groups are becoming widespread, from German cities like Hamburg (Tsang and Schuetze, 2018) to various European cities (Behrmann, 2019), and cities around the world (Hooftman et al., 2018). Because of the emissions of combustion vehicles and its results like driving bans, courier, express, and parcel (CEP) service providers have to think about replacing combustion vehicles (KE Consult, 2020).

Furthermore, research shows the need of switching to alternative propulsion systems (Morganti and Browne, 2018) to reduce pollution, global warming, and dependency on oil (Liao et al., 2017). Studies suggest that switching to ELCVs can be a promising way to decarbonize the transportation system and reduce the last-mile environmental impact (Iwan et al., 2021; Roca-Puigros et al; Tsakalidis et al., 2020). Hereby, focusing on charging renewable energies will be an important aspect (Juan et al., 2016).

Considering the sustainability of using electric vehicles for the last mile, this paper refers to its multidimensionality and hereby to the triple bottom line (TBL) model, considering sustainability from ecological, economic, and social perspectives (Elkington and Rowlands, 1999). Researchers have already investigated the dimensions of the TBL in the context of urban freight transport (Mello Bandeira et al., 2019) and highlighted the importance of investigating the TBL model in the context of sustainable urban last-mile logistics (Silva et al., 2023).

The CEP sector in Germany is characterized by only a few CEP service providers being responsible for 99 % of parcel deliveries (Meitinger, 2021) and being partly supported by more than 4,000 subcontractors (KE-Consult, 2023). In recent years, Amazon has also pushed delivery through its own logistics centers (About Amazon Team, 2020). Parcels are sorted for delivery via central and regional distribution centers of CEP service providers and subcontractors in order to handle them in less than 72 hours (Kuchenbecker et al., 2021). Although many CEP service providers have already started switching to alternative vehicles (Dachser, 2018) and thousands of electric vehicles have been ordered for fleets in the logistics industry (Tomlinson, 2020), 95% of the 13 million parcels are delivered every day with vehicles running on fossil fuels (Bernecker et al., 2022; Leerkamp et al., 2022). For new commercial vans in the first half of 2023, diesel vehicles account for 85.4% of new registrations and electric vehicles for only 7.1%. Compared to the first half of 2022, this is an increase of just under 2% in electric vans, and compared to the EU 2023, Germany is on average (acea, 2023). The tour profile in Germany is both urban and rural, with 50% of tours being a maximum of 50 km and 75% of tours being a maximum of 80 km (Schomakers et al., 2022). Some of the delivery services also rely on ELCVs from various suppliers or from their own development and achieve ranges of over 300 km with these (Ford Media Center, 2023). Companies being responsible for last-mile delivery belong to the level of CEP service providers or the level of subcontractors. In addition to these companies' managers, drivers also play a decisive role in last-mile delivery, finally being responsible for the deliveries and using the ELCV (Ehrler and Hebes, 2012), thus representing a third level.

Previous research has also demonstrated the importance of considering different user groups and their varied expectations and attitudes. For example, private consumers (Kim et al., 2023), last-mile operators of small and large companies (Lebeau et al., 2016; Morganti and Browne, 2018), and drivers (Wolff and Madlener, 2019), as well as CEP service provider drivers and shift managers (Ehrler and Hebes, 2012), have been researched. However, most of these studies only focused on one user group and did not focus on comparing these different user groups. Consequently, these studies didn't explore the varying attitudes of CEP service providers, subcontractors, and drivers towards the sustainability of electromobility. To devise comprehensive solutions, meet user expectations, and ensure the long-term viability of electric last-mile vehicles, examining these contrasting and converging viewpoints is essential (Hardman et al., 2018; Liu et al., 2019a). This study aims to address this research gap investigating the acceptance and attitudes of the logistics sector's various user levels toward the use of electric last-mile delivery vehicles. The study employs the German market as a representative example for European markets. In-depth interviews with CEP service provider managers, service providers, and drivers were conducted to address the following research questions:

How do logistics service providers evaluate the use of electric vehicles in the last mile from environmental, economic, and social perspectives?

Which contrasting or concurrent perspectives can be identified in this context between CEP service providers, subcontractors, and drivers?

We contribute to literature and practice in different ways. We were able to show the value of a multidimensional approach for investigating the last-mile sustainability. In addition, we were able to show the added value of expanding ecological, economic, and social sustainability to include technical process-related aspects. Furthermore, we identified the purchase price and range of ELCVs and the availability of public charging infrastructure as major barriers. Another contribution of our study is that the three levels of CEP—service providers, subcontractors, and drivers—have similar attitudes in many respects, but also different attitudes in some cases, for example the awareness of the environmental and social sustainability of last-mile electrification. While previous researchers have mainly analyzed the suitability of last-mile electrification in general, this study evaluates electrification from an internal perspective.

3.6.2 Theoretical Background

3.6.2.1 The Electrification of Freight Transport

Previous research on the electrification of freight transport focused on economic and ecological perspectives (Kaplan et al., 2016; Siragusa et al., 2022; Tsakalidis et al., 2020). For example, in their study in Berlin, Martins-Turner et al. (2020) demonstrated that a shift toward electrically powered trucks significantly reduces emissions, and Tsakalidis et al. (2020) calculated that switching to ELCVs in the European Union (EU) saves up to 30% of CO_2 emissions compared to previous light commercial vehicles. This shift produces other positive side effects, such as improved air quality (Tsakalidis et al., 2020). Moreover, the authors have shown that smaller ELCVs' costs are already more competitive than their counterparts with internal combustion engines. The cost of ownership for medium and large vehicles is still slightly higher; however, EU government incentives compensate for the difference (Tsakalidis et al., 2020). Siragusa et al. (2022) also compared the use of electric and internal combustion engine vehicles in the logistics sector from economic and environmental perspectives. Researchers have shown that electric vehicles are more environmentally friendly and lead to a reduction in emissions in small and especially large delivery areas. Furthermore, the costs are balanced over a time horizon of eight years, as lower operational costs compensate for the high purchase costs (Siragusa et al., 2022).

Researchers have also studied the transformation of the energy sector, including the electrification of transportation, in various scenarios. For example, a relatively slow, complete switch of the transportation sector to electromobility by 2070 is included in Shell's scenarios (Shell, 2021). In contrast, the International Energy Agency's "Net Zero by 2050" roadmap includes the more rapid decarbonization of transportation and thereby considers acceptance from technical, economic, and social perspectives (IEA, 2021). In all scenarios, the use of renewable energy plays a decisive role, which was also highlighted by Lindstad et al. (2023) in the context of decarbonization in road transport and by several studies in the context of switching to electromobility (Globisch et al., 2013; Juan et al., 2016).

Many studies have looked at the last-mile transition to electromobility in general, but fewer researchers have focused on the attitudes and acceptance of companies in the transportation sector in this context. However, the shift to last-mile electric vehicles also presents challenges for companies that must be considered. By interviewing last-mile operators and policymakers in European capitals, Morganti and Browne (2018) have shown that acceptance of electric vehicles is affected by concerns surrounding battery range and charging problems.

Anosike et al. (2023) and Noel et al. (2020) also identified technical barriers, such as range limitations and charging times, as key challenges. In contrast, other researchers have conducted real-life experiments to demonstrate that the range of electric vehicles is already sufficient for last-mile delivery (Iwan et al., 2021). Participants in the last mile consider the costs of vehicles and the installation of charging stations to be additional challenges (Anosike et al., 2023; Bosona, 2020). With no second-hand market for ELCVs, the costs are challenging and serve as a deterrent, especially for small companies (Tsakalidis et al., 2020). Various studies have investigated different user groups' acceptance of the last mile. Table 3.22 provides an overview of various empirical studies dealing with the acceptance of the use of electromobility for the last mile. The level considered in the studies is also listed.

Table 3.22 Excerpt of Relevant Empirical Studies Regarding the Acceptance of Switching to Electromobility

Study	Methodology	Company	User Group	Location
Lebeau et al. (2016)	Conjoint-based choice	Mainly small transport companies	Last-mile operators	Brussel
Morganti and Browne (2018)	Experiment-oriented interviews (N = 15)	Small and large companies	Last-mile operators and policymakers	France and the United Kingdom
Wolff and Madlener (2019)	Cross-sectional survey (N = 66)	CEP service provider	Drivers	Germany
Ehrler and Hebes (2012)	Interviews, questionnaires, logged routes, and observations	CEP service provider and freight forwarder	Drivers, shift managers, customers, and neighbors	Berlin
Noel et al. (2020)	Expert interviews (N = 227)	Different institutions	Experts	Denmark, Finland, Iceland, Norway, and Sweden
Iwan et al. (2021)	Real-life experiment with electric van	CEP service provider	Last-mile operators	Poland
Sierzchula (2014)	Interviews and pilot project reports	Public and private organizations	Fleet manager	The United States and the Netherlands

Table 3.22 shows that previous studies mostly address only one user group. Although Ehrler and Hebes (2012) deal with different stakeholders, the study was conducted more than 10 years ago and only investigated the area of Berlin. Furthermore, the focus of the study is not on comparing these user groups. Summing up, the presented studies did not highlight similarities or differences regarding the acceptance of electric vehicles of these groups and did not consider implications including the attitude of these different user groups. Accordingly, our study differs significantly from previous research by looking at and comparing different user groups in this context.

Another problem for logistics service providers is the weight limit for vehicles specified in the EU driving license, which restricts the volume of parcels because of the heaviness of the batteries used in electric vehicles (Tsakalidis et al., 2020). Moreover, charging electric vehicles with renewable energies is important for participant acceptance (Globisch et al., 2013; Juan et al., 2016).

In general, charging and charging infrastructure play key roles in CEP service providers' switch to electric delivery vehicles. For example, there are different requirements for grid connectivity, as the simultaneous charging of multiple vehicles within a fleet results in high electricity consumption with high peak loads, requiring optimal grid connectivity (Daina et al., 2017; Juan et al., 2016; Morganti and Browne, 2018). Previous research has shown that existing charging infrastructure is a particularly critical factor in the transition to electric vehicles because of their limited range compared to traditional delivery vehicles (Juan et al., 2016). Therefore, ensuring and maintaining effective and efficient charging infrastructure plays a crucial role in the acceptance and use of electromobility (Hardman et al., 2018; Sierzchula, 2014). Furthermore, access to charging infrastructure is one of the most important factors for predicting increasing electric vehicle sales (Mersky et al., 2016).

3.6.2.2 Economic, Environmental and Social Sustainability

Sustainability plays an increasingly important role for companies and society in general and is also a key consideration in the switch to electric vehicles in the last mile. However, sustainability should not be limited to environmental aspects; factors such as operational and social performance should also be considered (Bruzzone et al., 2021). In 2007, the Commission of the European Communities considered the sustainable development of urban mobility in the EU and distinguished between the environmental, economic, social, and health aspects (European Commission, 2007). Other studies have examined various aspects of sustainability in the logistics sector and especially the last mile. Bosona (2020) has highlighted the importance of cost efficiency, customer orientation, and environmental friendliness in last-mile logistics and further stated that environmental sustainability is lacking, and Silva et al. (2023) cited economic sustainability as the most important aspect for evaluating new solutions in urban last-mile logistics. Moreover, a study by Jiang et al. (2023) developed a systematic framework in the context of sustainable urban road alignment, including cost, social, and environmental factors.

Elkington and Rowlands‘ TBL model (1999) is an established approach used for this purpose, as it subdivides sustainability into three dimensions: economic, environmental, and social sustainability. In addition to companies' bottom lines and thus their economic growth, the TBL approach includes environmental sustainability and respectful treatment of the social environment (Elkington, 1994; Savitz, 2013). In this way, the TBL approach considers the whole of the company's responsibility (Jamali, 2006) and encompasses the sustainability perspectives discussed in urban last-mile logistics (Jiang et al., 2023). According to the TBL model, all three perspectives have to be taken into account equally when considering the sustainability of a company, product, or other approach (Elkington and Rowlands, 1999). The approach provides a robust basis for exploring and understanding sustainable concepts' acceptance factors and has successfully been applied in various contexts. For example, Pan et al. (2021) have divided corporate social responsibility into financial, social, and environmental dimensions to investigate corporate social responsibility in eco-innovation. Moreover, the principles of TBL have been considered in the context of sustainable construction (Goh et al., 2020) and Industry 4.0 (Khan et al., 2021). Khan et al. (2021) included four other literature reviews, investigating the TBL model in a technological context, and the approach has also been applied in logistics (Nikolaou et al., 2013). In summary, research has shown that ecological, economic, and social aspects should be considered for the long-term and sustainable establishment of ELCVs. Furthermore, users' attitudes and acceptance are essential to ELCVs' adoption. Studies have already identified various concerns participants have about the last mile (Morganti and Browne, 2018), which vary depending on the user level (Lebeau et al., 2016; Wolff and Madlener, 2019). Thus, differentiated research of CEP service providers, subcontractors, and drivers is important to obtaining a holistic overview and identifying commonalities and differences.

3.6.3 Methodology

We conducted a qualitative study to explore the last-mile electrification from an internal perspective and get unaffected statements from the different stakeholder of this sector. Qualitative research delves into human experiences, motivations, and behaviors by often employing smaller sample sizes, enabling a more intensive, detailed, and contextual exploration (Creswell, 2013). We wanted to generate both basic and more specific statements and subjective and individual attitudes. In line with Ehrler and Hebes (2012), Morganti and Browne (2018), and Noel et al. (2020) who investigated the acceptance of last-mile electrification of different user groups, semi-structured interviews were considered to be a suitable data collecting method for capturing a profound comprehension of the underlying perspectives, motivations, and complexities associated with ELCVs. Furthermore, qualitative research was conducted to obtain new and empirically valid results (Eisenhardt, 1989), offering great generalizability (Kaplan and Maxwell, 2005; Myers and Avison, 1997). A semi-structured guideline (Mayer, 2013) was developed first. This guideline was based on the literature presented earlier in this paper and in particular on the research of Wikström et al. (2016) and Tsakalidis et al. (2020). The researchers differentiated between various user groups, divided into fleet owners, fleet managers, and individual drivers, and examined the acceptance of electric vehicles in interviews with reference to range, battery capacity, individual driving behavior, and the deployment of vehicles in fleets. A semi-structured guideline enabled the comparison and evaluation of the interviews (Mayer, 2013), as this study involves different levels of participants. The interview questions were general and focused on the experience and expectations of electric vehicles and charging infrastructure in the last mile. Moreover, the interviews included questions about current processes. If the interviewees did not address the sustainability aspects, the interviewers asked about them specifically. The study used a general approach to avoid biasing the answers regarding the three dimensions (ecological, economic, and social) and to derive the dimensions' priorities.

The study was conducted in Germany, within the context of the German CEP sector. This sector has been witnessing an increasing consolidation, with a scant number of leading logistics service providers handling 99% of domestic parcel shipments (Meitinger, 2021). The German market presents a conducive environment for exploring fleet electrification given its dense urban infrastructure, high frequency of deliveries, and standardized routes (Schomakers et al., 2022). Additionally, the collaborative endeavors between CEP service providers and subcontractors concerning delivery services are driven by various factors including the management of peak periods, geographic coverage, and enhancing flexibility in delivery service provision. Given the attributes of the German CEP sector, a methodologically rigorous approach was adopted for the interview process aimed to gaining deep insights. The process involved engaging employees from CEP service providers and subcontractors, as well as drivers, striving to include as many participants as feasible to ensure a broad spectrum of perspectives. The interviews encompassed six employees from CEP service providers, ten employees from subcontractor management teams, and ten drivers.

The engagement with subcontractors and drivers revealed the challenge of accessibility owing to their deep involvement in daily operational activities. A systematic approach was employed in the interview process, where the addition of participants continued until a point of data saturation was reached. Data saturation is deemed to occur when subsequent interviews cease to unveil new insights or information pertinent to the research questions (Hennink and Kaiser, 2022). Our sampling procedure ensured the adequacy and appropriateness of the sample size, while maintaining the integrity and depth of the qualitative inquiry. As more interviews were conducted, it was discerned that the generation of new information diminished, affirming that data saturation had been achieved, thereby substantiating the appropriateness of the sample size for this study.

Most of the representatives of the CEP service providers are actively involved in the companies' (sustainable) strategic orientation. The subcontractors' employees are part of their companies' management teams and are actively involved in sustainable decision-making processes. Heterogeneous distribution with regard to previous experience with electric vehicles was guaranteed for the selection of drivers as interview participants (IPs). Overall, a heterogeneous group of IPs was chosen, and their characteristics are listed in Table 3.23. To exclude the influence of gender, the study only included male participants, which illustrates the high number of male employees in this sector, as only 20% of employees in the logistics, transportation, and traffic sectors in Germany are female (BVL and transport logistic, 2019).

Table 3.23 Overview of IPs

IP	User group	Job	Use of electric vehicles	Delivery area
IP1	CEP service provider	Central logistic manager	Increasingly thinking about the use of electric vehicles	Urban and rural
IP2	CEP service provider	Clean operations manager	Uses electromobility widely	Urban and rural with longer driving cycles

(continued)

Table 3.23 (continued)

IP	User group	Job	Use of electric vehicles	Delivery area
IP3	CEP service provider	Sustainability and city center logistics manager	Uses electric vehicles sporadically	Urban and rural
IP4	CEP service provider	Department of Strategy	Has been using electric vehicles for a few years	Urban and rural
IP5	CEP service provider	Connecter of internal and external stakeholders	Has been using electric and hybrid vehicles for many years	Urban and rural
IP6	CEP service provider	Finance and supply manager	Deployed and developed various concepts for using electric vehicles	Urban and rural
IP7	Subcontractor (courier, parcel, and express)	Logistics network chairman	Has contact with companies testing the use of electric vehicles	Urban and rural
IP8	Subcontractor (courier, parcel, and express)	Authorized representative and shareholder	Already tested electric vehicles but not using currently	Urban and rural
IP9	Subcontractor (courier and overnight)	Managing director and additional branch manager of a logistics network	Has been using electric vehicles for many years but not using currently	Focus on rural
IP10	Subcontractor (courier)	Advisor to management	Sporadically uses electric vehicles	Focus on urban
IP11	Subcontractor (courier, parcel, and express)	Advisor to management	Has tested several electric vehicles	Urban and rural
IP12	Subcontractor (parcel and general cargo)	Managing director	Has tested several electric vehicles and currently uses electric bikes	Focus on urban
IP13	Subcontractor (parcel)	Independent entrepreneur	Uses electric vehicles	Focus on rural
IP14	Subcontractor (courier and parcel)	Fleet manager and electric vehicle coordinator	Uses electric vehicles	Urban and rural
IP15	Subcontractor (parcel)	Dispatcher	Has experience using electric vehicles	Focus on urban

(continued)

Table 3.23 (continued)

IP	User group	Job	Use of electric vehicles	Delivery area
IP16	Subcontractor (courier, parcel, and express)	Managing director	Has extensively considered the use of electric vehicles	Urban and rural
IP17	Driver (courier)	Driver for CEP service provider	Has used an electric vehicle for several months	Urban
IP18	Driver (parcel)	Driver for subcontractor	Uses electric vehicle	Urban and rural
IP19	Driver (parcel and courier)	Independent courier driver for subcontractor	Has used an electric vehicle for six years	Urban and rural
IP20	Driver (express)	Driver for subcontractor	Does not use an electric vehicle but gained experience through conversations with colleagues	Focus on rural
IP21	Driver (courier and parcel)	Driver for CEP service provider	Has used an electric vehicle for ten years	Urban
IP22	Driver (courier)	Driver for subcontractor	Does not use an electric vehicle but gained experience through conversations with colleagues	Focus on urban
IP23	Driver (courier)	Driver for subcontractor	No experience	Urban and rural
IP24	Driver (express)	Driver for subcontractor	Does not use an electric vehicle but gained experience through conversations with colleagues	Focus on urban
IP25	Driver (parcel)	Charging park operation manager and part-time driver for a CEP service provider	Has experience using electric vehicles	Focus on urban
IP26	Driver (parcel)	Former charging park area manager and part-time driver for a subcontractor	Has experience using electric vehicles	Focus on urban

A total of 26 in-depth interviews were conducted between April 2021 and February 2022. The interviews were conducted partly on site and partly digitally because of the COVID-19 pandemic and the logistical effort required. The interviews were recorded, anonymously transcribed, coded, and subsequently analyzed using MAXQDA software. The process was based on Mayring and Fenzl's approach (2019) to content analysis, whereby the researchers formed the main categories following a deductive approach based on the TBL model and formed the subcategories following an inductive approach when analyzing the transcripts. Furthermore, the researchers revised the code system through a peer review process after the first seven interviews. The coding was carried out independently by two researchers and subsequently discussed and merged (Mayring, 1994). Table 3.24 shows an overview of the category system.

Table 3.24 Category System

Main Category	Subcategory
Ecological sustainability	CO_2 emissions
	Batteries
	Green electricity
	Alternative drives
Economic sustainability	Electric vehicle prices
	The availability of electric vehicles
	The requirements of building charging infrastructure
	The availability of charging infrastructure
	The price of electricity
	Financial support
	Competition
Social sustainability	The urban landscape
	Legal requirements
	Image
	Social responsibility
Technical process-related aspects	The range of electric vehicles
	Technical characteristics of electric vehicles
	Changes of processes
	Test purposes

As the main categories are based on the TBL approach, they include ecological, economic, and social sustainability. However, some of the IPs' statements cannot easily be assigned to one of the three main categories; instead, they provide accompanying and correlating technical and process-related aspects. Therefore, a fourth main category, technical process-related aspects, was added. Four subcategories were assigned to ecological sustainability, seven subcategories were assigned to economic sustainability, four subcategories were assigned to social sustainability, and four subcategories were assigned to technical process-related aspects. Subcategories related to two or more main categories were finally assigned to the main category, and the greatest agreement in terms of context was identified in the interviews.

The operationalization of the TBL approach has not been standardized yet; thus, several approaches, such as the approach used by Schulz and Flanigan (2016) were developed. However, no approach was established for the long term; therefore, the inductively formed subcategories are based on former literature and the interviews' characteristics in this study.

3.6.4 Results

The number of subcodes indicates an interesting result, showing that the economic perspective has been discussed most extensively and intensively by the IPs. This is also illustrated by the number of coded text passages (ecological sustainability: 33, economic sustainability: 178, social sustainability: 53, technical process-related aspects: 60), showing the dominance of economic sustainability in the IPs' statements. However, in contrast to ecological and social sustainability, economic sustainability tends to be viewed negatively for electromobility in the last mile during the interviews. In addition, some of the technical process-related aspects are viewed positively and others negatively.

3.6.4.1 Ecological Sustainability

Within the main category of ecological sustainability, the IPs discussed CO_2 emissions most intensively. Accordingly, for CEP service providers, reducing CO_2 use is a decisive factor when switching to electromobility and is also linked to companies' goals. Although managers of subcontractors also pointed out the positive effect of CO_2 savings, they did not highlight this aspect, and fewer Ips of this level mentioned these savings. Nevertheless, they expressed their company's intention to drive using less CO_2 or even be CO_2 neutral in the long term and to achieve these goals by using electric mobility (cf. IP12). Moreover, private consumers' expectations of CO_2-neutral deliveries were given as a motivation (IP10). Almost all IPs at the driver level mentioned environmental protection as one of the main reasons for switching to electric vehicles.

However, only one CEP service provider manager and one subcontractor manager specifically addressed the importance of renewable electricity as a means to reduce emissions in the long term (IP1, IP14). Only one subcontractor addressed the environmental sustainability of electric vehicles after they have been used (including waste disposal and especially battery disposal) (IP14).

The importance and benefits of other alternative propulsion technologies, such as hydrogen vehicles, were only mentioned by a few interviewees. However, according to those interviewees, these technologies should play a central role in the logistics sector's long-term ecological sustainability (cf. IP1). No differences between the levels of IPs were recognized for this subcategory. The interviews revealed a contradiction between the logistics companies' intentions and behaviors. The reduction of emissions was stated as a goal of the companies; however, when it comes to the decision to use electric vehicles, specifically electric fleets, not all IPs seem to have intensively focused on environmental sustainability.

3.6.4.2 Economic Sustainability

Ecological sustainability is often emphasized in the context of sustainability. However, this study shows that IPs are particularly influenced by economic aspects when discussing the switch towards electromobility. Thus, the number of categories and coded text passages indicate that economic sustainability is the decisive dimension for those considering the switch to electric vehicles. Hereby, especially the negative aspects of economic sustainability in this context are highlighted by the IPs. Electric vehicles' high purchase prices were mentioned by three managers of a CEP service provider and described as a limiting factor (cf. IP3). This aspect was also mentioned by some subcontractors and self-employed couriers who cannot afford electric vehicles (cf. IP23). However, all levels mentioned the high purchase price and did not consider the total cost of ownership during the interviews.

In addition to the price, electric vehicles' lack of availability was discussed by most of the CEP service providers and some of the subcontractors and drivers, as many vehicles do not meet the requirements for usage in the logistics industry (cf. IP4). Moreover, long delivery times were criticized (cf. IP7), and the lack of second-hand electric vehicles was mentioned by drivers in particular (cf. IP23).

Additionally, the importance of owning charging infrastructure was mentioned in the interviews. However, the requirements for the construction of charging infrastructure are a significant barrier when switching to electric fleets and were discussed by managers of CEP service providers in particular. This level of IP highlighted the high costs of a suitable location, an appropriate number of charging points, and the power supply (for example, transformer stations) (cf. IP1). Regarding the charging process, long overnight charging times with 3.7 kilowatts were noted as being feasible (cf. IP2). Subcontractors also reported problems setting up charging infrastructure, including a lack of permission (cf. IP11), long wait times to repair faults (cf. IP9), and general difficulties for smaller companies (cf. IP7).

Furthermore, all IP levels criticized the low availability of public charging infrastructure, which makes the switch to electric vehicles difficult (cf. IP6). According to the IPs, delivery routes without the opportunity for public charging are too unsecure (cf. IP 17). Some IPs mentioned the price of electricity and talked about how diesel and gasoline are cheaper in the long run (cf. IP14), which is highly relevant to the conversion of entire fleets (cf. IP7). This aspect was highlighted by subcontractors in particular.

Two managers of CEP service providers and two subcontractors raised the relevance of governmental support. They pointed out the importance of promoting infrastructure (cf. IP5), vehicles (cf. IP13), and cargo bikes (cf. IP10) for a sustainable transition to electromobility. Only two managers of CEP service providers and one system partner felt pressured to switch to electric mobility by their competitors' activities (cf. IP3).

3.6.4.3 Social Sustainability

The IPs mentioned several aspects of social sustainability. Three CEP service provider managers mentioned that electric fleets and charging locations could reduce traffic in cities and improve air quality, thus offering advantages for society (cf. IP3). In contrast, subcontractors considered the use of electric fleets neutral with regard to the urban landscape and expected no changes (cf. IP7), focusing more on the influence of private users' electrification (cf. IP16). However, Berlin was highlighted positively because of the number of cargo bikes (cf. IP10). Most drivers mentioned the positive impact of electric fleets on the urban landscape, with vehicles being much quieter at a reasonable speed (cf. IP26). Furthermore, five CEP service provider managers and four subcontractors addressed the importance of switching to electric fleets for the last mile because of legal enforcements, including driving bans for combustion vehicles in cities (cf. IP2), CO_2 taxation (cf. IP11), and special permits for electric vehicles (cf. IP9). However, CEP service provider managers' statements about legal restrictions were more specific than those of subcontractors' managers, and drivers did not address this point. The public image of electric vehicles in the logistics sector, which has a high media impact, was also mentioned by the IPs, including six subcontractors, two drivers, and one CEP service provider. This issue was almost exclusively addressed by interviewees who had already actively used electric vehicles. For example, they pointed out the low sound level of electric vehicles (cf. IP21). In contrast, one IP noted that the use of electromobility was just a lighthouse project (cf. IP7). The last category of the social sustainability dimension, social responsibility, summarizes the extent to which the interviewees were aware of their responsibility toward the public to switch to electric fleets. These aspects were only addressed by the CEP service providers, pointing out the importance of social responsibility of companies (cf. IP 6).

3.6.4.4 Technical Process-Related Aspects

Several aspects highlighted by the interviewees could not be directly assigned to one of the three sustainability categories; instead, they accompanied and correlated with sustainability. Therefore, they were assigned to the main category of technical process-related aspects. One of the most frequently discussed aspects during the interviews was the range of electric vehicles. This characteristic was assigned to technical and process-related aspects and is essential for the sustainable implementation of electric vehicles, according to the interviewees. Sixteen of the 26 IPs, including CEP service provider and subcontractor managers and drivers, identified the range of vehicles as a major barrier for a partial or complete switch to electromobility. They criticized the range of the batteries and additionally pointed out limitations based on weather or traffic conditions (cf. IP14). Interviewees at all three levels also stated that the range is currently insufficient, especially for rural trips (cf. IP4). In this context, a manager from a CEP service provider also mentioned that different vehicle sizes have to be considered for delivery, which also influences the vehicles' range (cf. IP5).

In contrast, the overall attitude toward electric vehicles' technical differences was positive. Active users of electric vehicles (drivers in particular) emphasized the pleasant driving experience and the less stressful delivery procedure without ignition processes or the use of the clutch (cf. IP17). Drivers' positive experiences were also highlighted by two managers of CEP service providers, who highlighted the advantage of being less susceptible to failure (cf. IP3). Moreover, subcontractors also pointed out electric vehicles' lower maintenance requirements (cf. IP10). However, electric vehicles' heavy weight was also mentioned by one subcontractor as a technical problem when loading vehicles (cf. IP11).

Furthermore, the use of electric vehicles for the last mile includes changes to process sequences.

According to the CEP service providers, the longer charging times do not affect drivers (cf. IP6). However, according to the subcontractors, this depends on the agreements that the companies have with their drivers, who are no longer able to take the delivery cars home overnight because of charging processes (cf. IP15). Working processes in general have to be adapted because of charging durations and locations as well as range (cf. IP11). Subcontractor managers' statements regarding this subcategory were more consistent than those of CEP service providers. Drivers pointed out the influence of heating on the driving range (cf. IP17), expensive repair costs for those who are self-employed (cf. IP21), and the need to learn about different charging and payment systems (cf. IP22) when using electric vehicles.

3.6.4.5 Framework

Fig. 3.19 represents a framework derived from the results of the qualitative study. The framework comprises aspects of freight transport's sustainable electrification. Some statements (such as those involving range) were difficult to assign to a category related to Elkington's TBL model (1994). Therefore, technical process-related aspects complement the model. The framework does not include correlations of effects and represents a graphical illustration of the results.

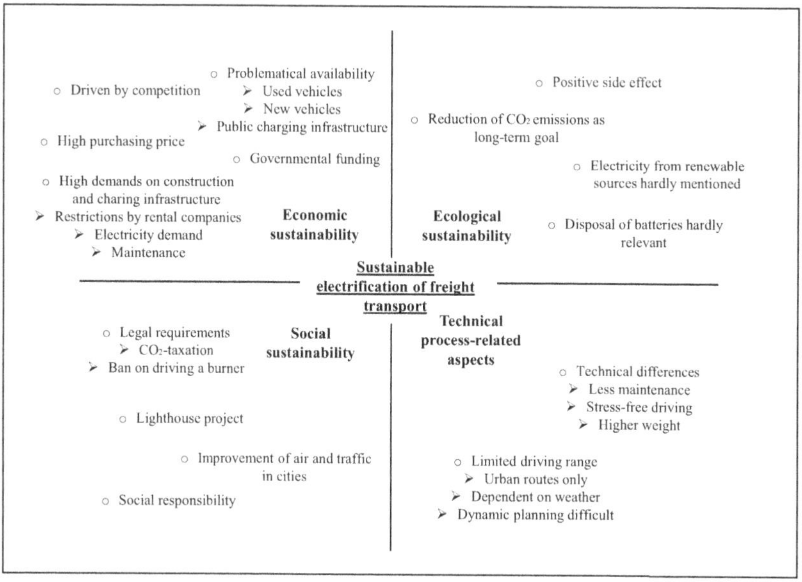

Fig. 3.19 Framework for the Sustainable Electrification of Freight Transport

3.6.5 Discussion

Switching to electric vehicles is a key response to reducing CO_2 emissions and addressing climate change (Siragusa et al., 2022; Tsakalidis et al., 2020). Although the reduction of CO_2 emissions was briefly mentioned during the interviews by CEP service providers and drivers, the actual transition to large-scale fleet electrification lags behind environmental sustainability goals. The importance of renewable energy (Lindstad et al., 2023) in this context was also hardly addressed by the interviewees. Previous studies have highlighted the importance of electric bikes for decarbonization (Diaz-Ramirez et al., 2023), but the interviews have shown that they are only suitable for small courier trips and not for larger parcel delivery, especially in mountainous regions.

From the interviewees' perspectives, economic considerations played a pivotal role in last-mile electrification, especially for fleet managers. As already discussed in the literature (Anosike et al., 2023), high purchase prices and charging infrastructure costs were highlighted as limiting factors, despite evidence suggesting that ELCVs are cost-competitive in terms of total costs of ownership (Lal et al., 2023; Tsakalidis et al., 2020). Moreover, the construction of larger charging parks and the opportunities of buying ELCVs were identified as challenges by CEP service providers in the context of economic sustainability. Furthermore, public charging infrastructure was deemed important, consistent with previous research (Juan et al., 2016).

In the context of social sustainability and the societal impact of electrification, CEP service providers and drivers as well as previous literature noted improvements in air quality and congestion in cities (Quak et al., 2016). Furthermore, last-mile freight transport has a negative impact on noise and congestion in cities. During the interviews, the drivers highlighted electric vehicles' positive characteristics, such as their quietness and their positive effects on the cityscape. CEP service providers also pointed out positive side effects on cities and society, whereas subcontractors evaluated these effects as neutral but pointed out the advantages of an improved public image resulting from ELCV use. The interviewees and CEP service provider in particular also stated the driving force of legal requirements behind last-mile electrification (Braun, 2019; Hooftman et al., 2018).

Previous studies covered further sustainability aspect, including operational performance (Bruzzone et al., 2021), traffic, and developing and engineering factors (Jiang et al., 2023). During the interviews, several aspects could not be clearly assigned to one of the three categories adapted from the TBL model (Elkington, 1994); therefore, a fourth category was established for the technical process-related aspects. Range limitations were a recurring concern for interviewees across all levels, particularly for rural deliveries. The interview findings aligned with previous research of Morganti and Browne (2018) and Noel et al. (2020). However, Iwan et al. (2021) have demonstrated through a real-life experiment that the range of ELCVs is already competitive for the last mile and Tsakalidis et al. (2020) pointed out companies being able to make rational decisions regarding delivery route distances. In addition to uncertainties surrounding range, other concerns were mentioned, including the fact that vehicles cannot be left at the driver's home overnight because of charging limits and uncertainties regarding payment systems and the limits that batteries' additional weight imposes (Tsakalidis et al., 2020). However, drivers also highlighted the pleasant driving experience during the interviews.

Our study has shown that CEP service provider managers, subcontractors, and drivers partly have different attitudes regarding last-mile electrification and have already had different experiences with ELCVs. Interestingly, all three levels identify similar major barriers for a large-scale electrification. These include the high purchase prices of ELCV and charging infrastructure, the limited vehicle range, and limited opportunities for public charging. Thus, especially the economic sustainability and technical process-related aspects are still perceived negatively. Moreover, some aspects differed among the groups and are illustrated in Fig. 3.20. Notably, CEP service providers and drivers express their interest in the positive environmental aspects of ELCVs. In contrast, subcontractors primarily base their electrification decisions on economic considerations, which also influence the decisions of the other two groups. However, CEP service providers and drivers also consider the positive impact of ELCVs on the urban environment and the enhanced driving experience of the drivers.

Theoretical and practical implications can be derived from this study. First, the use of the TBL model's (Elkington, 1994) subdivisions for last-mile electrification's sustainability is suitable, however, an additional subdivision for factors such as technical processes should also be considered. This was also demonstrated by previous researchers, such as Jiang et al. (2023), who considered developing and engineering factors. Furthermore, it has been shown that a breakdown of the last mile into different levels can generate added value for research and practice. The three levels exhibit varying stances concerning ecological, economic, and social sustainability as well as technical process-related aspects in the context of last-mile electrification. Notably, active usage of ELCVs tends to foster

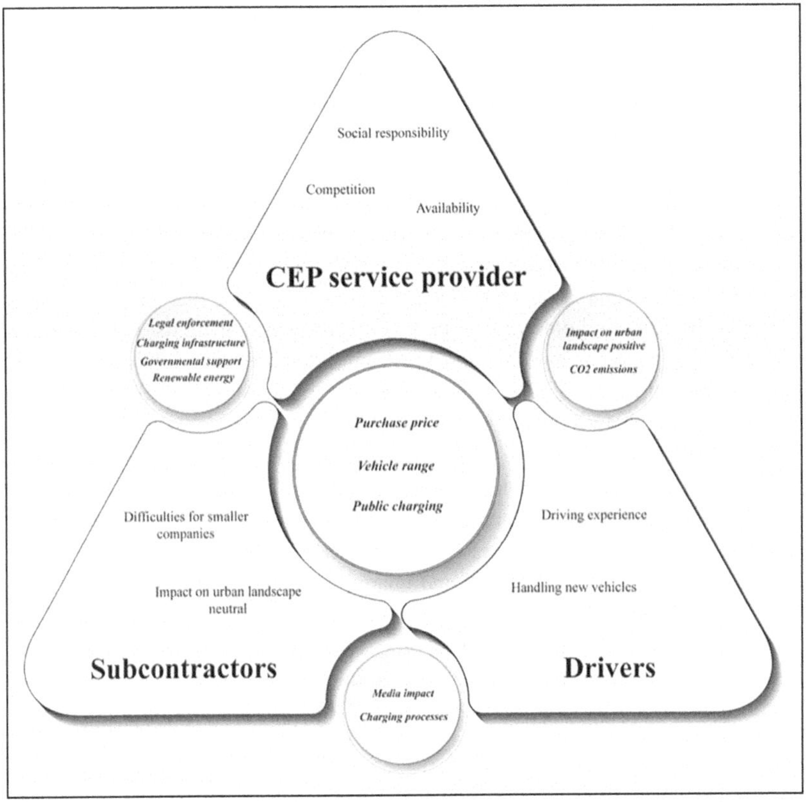

Fig. 3.20 Interdependencies between the Three Levels. (Note: Aspects mentioned by all three levels are depicted in the center of the figure, aspects mentioned by two level are allocated to the related arrow and aspects mentioned only by one level are attributed directly to them)

distinct perspectives and establish a more favorable disposition toward these vehicles. Because of this, an ongoing exchange of ideas and experiences among these three levels, as well as between experienced and unexperienced users of ELCVs, is important to facilitate the large-scale adoption of electrification. In particular drivers introduce unique viewpoints on certain aspects that should be pivotal in discussions involving managers, policymakers, and other stakeholders.

To achieve a successful last-mile electrification in the future, it is important to transparently address the uncertainties pertaining to various facets of this transition. Specifically, with regard to the economic sustainability and technical process-related aspects of ELCVs, the expectations of the IPs are misaligned with findings from previous studies and field experiments. Research conducted by Tsakalidis et al. (2020) has demonstrated that ELCVs are already cost-competitive with traditional combustion engines when considering the total cost of ownership. However, the IPs primarily focus on the initial purchase price when considering the purchase of ELCVs. Additionally, research by Iwan et al. (2021) illustrated that the range of ELCVs is well-suited for last-mile operations, contradicting the concerns raised by the IPs regarding the limited range. Since both the high purchase price and limited range have been identified as significant barriers for all levels, increased transparency is required regarding the unique characteristics and capabilities of ELCVs. Moreover, fostering enhanced communication within the industry is essential for addressing these misconceptions and promote the adoption of last-mile electrification. Thus, policymakers should continue to support the spread of charging infrastructure and, in cooperation with associations, for example, demonstrate ELCVs' cost-competitiveness (Lal et al., 2023) and the already sufficient range by providing test vehicles. This will also help overcome drivers' insecurities and demonstrate the positive characteristics of electric vehicles. The exchange within the industry between companies that have already gained experiences with ELCVs, and less experienced companies should also play a central role. Existing association structures, such as the BdKEP (BdKEP, 2023) and BIEK (BIEK, 2023) in Germany, the ELA (ELA, 2023) at European level, and other regional and international associations, should also be used for more transparency regarding the mentioned aspects. Overall, the focus of communication should not only include ecological and social benefits, but also address the economic sustainability and technical process-related aspects of ELCVs. Furthermore, governmental restrictions on internal combustion engines could be advocated for more strongly pushing the shift towards last-mile electrification (McLeod et al., 2020).

European car manufacturers should focus on the development and distribution of light commercial electric vehicles to promote the electrification of the last mile (Tsakalidis et al., 2020). The establishment of a second-hand market for vehicles will also play a crucial role from a last-mile perspective, as it will grant smaller companies access to electric vehicles (Tsakalidis et al., 2020). However, regional characteristics have to be taken into account for last-mile electrification. Thus, some regions' topology and infrastructure might be difficult to access by larger vehicles and cars (Comi and Savchenko, 2021), whereas others are not usable by cargo bikes, as seen in this study.

The insights derived from the IPs in Germany's last-mile logistics sector align closely with findings from other countries and are therefore applicable in an international context. In Denmark, Finland, Iceland, Norway, and Sweden Noel et al. (2020) identified limited vehicle range, high purchase prices, and insufficient public charging as the biggest barriers for using electromobility in the logistics sector. Similar observations were reported by Morganti and Browne (2018) in France and the United Kingdom, as well as by Lebeau et al. (2016) in Brussels. Moreover, these studies pointed out disadvantages of ELCVs being more valued than their advantages, and the attitude of active users of ELCVs being more positive. Finally, Sierzchula (2014) depicted similar results and circumstances in this context between Europe (the Netherlands) and the United States, making the results of this study even more applicable in an international context.

3.6.6 Conclusion

Our study addressed two research questions with the first question exploring how logistics service providers assess the adoption of electric vehicles in the last-mile delivery from environmental, ecological, and social perspectives. Findings revealed that economic sustainability played a central role in driving the transition to electromobility in this context, with environmental and social aspects often considered as secondary. Nonetheless, it has been demonstrated that the transition to electromobility in the last-mile context is yet to attain economic sustainability, even though it is perceived positively in terms on its environmental and social impacts. Furthermore, technical process-related aspects were identified as a fourth perspective, also being considered for the last mile's sustainable electrification. The second research question aimed to uncover contrasting or parallel viewpoints among CEP service providers, subcontractors, and drivers in this context. Most last-mile companies are aware of their social responsibility, so electrification should and can be further promoted with the support of the government, associations, and other stakeholders.

Several limitations have to be considered when examining the results of the study, leading to the need for further research. A qualitative study is suitable for gaining greater generalizability of a topic (Kaplan and Maxwell, 2005) but should be verified by additional quantitative studies. Furthermore, the survey was conducted in Germany and on the basis of German last-mile structures. Although the implications can be transferred to other countries, they should also be verified in other structures. The results were sorted into categories with reference to the TBL model, whereby the dimensions merge smoothly into each other for some aspects. Further dimensions should be taken into account in future research, as shown in this study, as well as research by Jiang et al. (2023). Research should also again compare last mile expectations with the current characteristics of ELCVs. The differences shown between research and last-mile stakeholder statements should encourage future research and transparent design. This study demonstrates the importance of understanding the participants of the last mile and shows that acceptance does not always correspond to scientific research for the evaluation of electrification.

4 General Conclusion

4.1 Research and Theoretical Implications

Technologies are omnipresent and influence the way in which individuals conduct their daily lives through the utilization of mobile devices and digital platforms and transform the way companies operate. A variety of factors can be identified that influence the acceptance and resistance of technologies. The objective of this dissertation was to examine a range of innovative technologies and the intention to use these technologies. Six different technologies were examined in six essays and its major results and theoretical implications as well as future research opportunities are illustrated in the following. Referring to the framework of this dissertation, determinants of acceptance and resistance based on the UTAUT (Venkatesh et al., 2003) and the IRT (Ram and Sheth, 1989) are derived from each essay.

Essay 1: Ethics and Morality in AI—A Systematic Literature Review and Future Research

In the first essay, a systematic overview of ethics and morality in the context of AI is presented. Seven fields into which this area can be subdivided are identified: ethical and moral implications of AI, approaches of ethical handling of AI, demand for governance and regulations, AI programming requirements, human moral judgment and moral responsibility, moral status of AI and importance of law and politics. Many of the systematized articles address ethical and moral implications and approaches to the ethical use of AI, particularly in the domains of healthcare (Brady and Neri, 2020) and business (Hamilton and Davison, 2021).

J. Selter, *Acceptance and Resistance of Innovative Technologies*, Handel und Internationales Marketing Retailing and International Marketing,
https://doi.org/10.1007/978-3-658-48594-8_4

Further research may also be directed towards areas such as the public sector and topics like education (Chounta et al., 2021). The effect of AI on the personalization of consumer applications (McAleenan, 2020) represents a promising avenue for future research on the ethical implications of AI for private users. Additionally, there is a growing body of literature that focuses on the development of principles for the responsible use of AI (Floridi et al., 2018). A holistic approach based on human rights, which can be implemented globally (Gibbons, 2021), might be a fruitful direction for future research in this context. To address the complexity of AI programming and the diverse application contexts in which it is used, future research may benefit from a pluralistic ethical approach (Gordon, 2021). The same applies to other pivotal domains, such as the political regulation of AI, where comprehensive approaches could be cultivated through research (Li, 2021). Legal accountability could be a central concern in AI programming, and the formulation of regulations for programming in research could prove crucial (Burton et al., 2020). The essay shows that trust in the ethical utilization of AI seems to be a fundamental element influencing the behavioral intention to employ this technology. Results allow a derivation of various determinants of acceptance and resistance. Factors such as ethical principles and regulations serve as determinants of acceptance. Furthermore, trust in the technology and transparency for consumers regarding aspects such as responsibility, moral status, and data usage of AI, are also crucial determinants. A comprehensive understanding of the ethical principles can enhance the intention to utilize the technology, representing a further determinant of acceptance. Conversely, ambiguities concerning moral accountability, and a lack of ethical benchmarks can be identified as determinants of resistance. Uncertainties regarding the responsibility gap can also be recognized as determinants of resistance. Essay one highlights that the acceptance and resistance of AI is profoundly associated with ethical and moral concerns, underscoring the necessity for future research to consider these factors in more detail.

Essay 2: Aspects Driving Customers' Intention to Use Automated Purchasing Processes

The second essay considers automated purchasing processes in the context of smart home applications. Findings indicate that consumers exhibit a preference for traditional shopping methods and a rejection of innovative and automated processes with which they are not familiar. Consumers are reluctant to adopt these technologies, even when presented with clear advantages such as time savings or convenience (Bayer et al., 2021). Instead, they are deterred by perceived disadvantages, including the effort required (Gupta et al., 2021b) and financial costs (Gonçalves et al., 2020). Previous research on key variables in this context, such as perceived trust (Macnish and Fernandez Inguanzo, 2019) and behavioral control (Jörling et al., 2019), was also affirmed by the first essay. Moreover, based on the results of essay two, the Uncanny Valley effect may be applicable to the context of automated purchasing processes, particularly in relation to the level of automation. Automated processes might be accepted by consumers until they reach a level of autonomy that provokes a sense of unease and is perceived as too autonomous (Boradbent, 2017; Mori et al., 2012). A further area of investigation might be the examination of different levels of automation, given the growing relevance and usability of fully autonomous processes (Kahlert et al., 2017). The investigation may concentrate on the Uncanny Valley effect, which postulates that robots that are excessively similar to human beings cease to appear familiar once a specific degree of similarity is attained (Mori et al., 2012). This phenomenon can also be applied to other contexts, where an excessive degree of autonomy in processes and technologies may also be rejected. Accordingly, further investigation could be conducted into the potential applications of autonomous technologies like autonomous shopping processes. The focus could be on ensuring consumer autonomy and consent, as well as transparency and accountability in these processes. The second essay offers insight into the various determinants of acceptance and resistance of technologies. One factor influencing acceptance in the context of automated shopping processes is the extent to which individuals perceive control over the process, including the ability to influence and make changes. Additionally, trust in the automated process, such as the reliability of the technology or data security represents a further determinant of acceptance. The loss of control over the process and discomfort with excessive autonomy of the technology can be identified as factors of resistance. In general, high autonomy represents a factor of resistance in this context.

Essay 3: Transferring Trust, Risk and Security on Users' Intention to Trade Cryptocurrency, Using PayPal as an Example

The third essay investigates the possibility of transferring trust and trust-related aspects from existing platforms to new technologies. Results demonstrate that the transfers of trust, risk, and security from a payment service provider to cryptocurrencies follow the expectations of the trust-transfer model (Stewart, 2003). This has various theoretical implications, as the trust-transfer model was originally developed for transfers between websites (Stewart, 2003, 2006). While the transfer of trust from service provider to payment option has been demonstrated in several other application contexts, including the transition from online to mobile payments (Cao et al., 2018; Lu et al., 2011), it has yet to be confirmed in this particular instance. As new innovative technologies are regularly developed, future research could investigate whether the trust-transfer model can also be applied to other linkages between technologies and existing systems and could represent a general approach in this context. In e-commerce, various connections might be examined here, such as users switching from established platforms like Amazon to newer ones. Additionally, research has highlighted that expanding the trust-transfer model to incorporate the associated factors of risk and security might offer significant benefits. This statement is in line with previous research, which primarily focused on trust transfer in the payment sector (Gong et al., 2020) and which emphasized perceived security and perceived risk as decisive determinants for the use of technologies and services (Bhatnagar and Ghose, 2004; Tandon et al., 2018). Accordingly, an extension of the established trust-transfer model with the additional determinants of security and risk appears to offer a valuable contribution to the existing body of knowledge. Further research might examine the transfer of additional attributes, including usability, efficiency, or sustainability. Results of the third essay demonstrate that in the context of cryptocurrencies, the transfer of trust, security, and risk perception seems to play a significant role regarding the intention to use them. In particular, security represents a central factor influencing behavioral usage intention. These findings have implications for the acceptance and resistance of cryptocurrencies and related technologies. The perceived security of a platform and the trust placed in it represent key determinants of acceptance, which can also be transferred to the technologies used via that platform. High perceived risks, for example concerning data disclosure, represent determinants of resistance. Similarly, a lack of trust and uncertainty are also relevant determinants of resistance, which can also be transferred to other technologies based on the trust transfer theory and thus negatively influence the intention to use these technologies.

Essay 4: Digital Nudging to Reduce Product Returns in E-Commerce: Investigating Consumers' Intentions and Environmental Implications

The fourth essay examines the potential of digital nudges as an innovative approach in online retailing to reduce returns and contribute to a more environmentally sustainable design of e-commerce. Results demonstrate that digital nudges seem to exert a considerable influence on consumers' decision-making processes and shape diverse behavioral patterns concerning environmentally conscious actions, purchases, and intentions to return and reuse. The framing nudges showed the most substantial impact in the essay and in literature (Berger et al., 2022). While the dissertation considers digital nudges individually, future research might investigate the potential for complementary effects or interference between digital nudges when they are used simultaneously or in combination. Furthermore, the results substantiate the effect postulated according to the goal-framing theory (Lindenberg and Steg, 2007). In consequence, the behavioral intentions of consumers are shaped by hedonic, normative, and gain goal frames, which are important for forming environmentally friendly behaviors. Consumers tend to prioritize their internal moral and ethical codes over social influences when making decisions about pro-environmental behavior in the context of mobile shopping. No effect of the gain goal frame on consumers' intention to return the product was identified, diverging from the findings of previous studies on sustainable fashion consumption (Lundblad and Davies, 2016). This discrepancy may result from consumers associating choice orders with unavoidable returns, thereby making the environmental consequences more salient. Digital nudges provide potential applications in several other contexts, including energy consumption, mobility, and recycling, offering further investigation. Results of the fourth essay illustrate that digital nudges can promote pro-environmental behavior. In the future, it may be feasible to quantify the actual impact of digital nudges on environmental and economic sustainability by numerical measurement. Regarding the ethical challenges of digital nudges, it is important to guarantee that the autonomy of consumers is not unduly influenced or violated. Examining relevant regulations can facilitate the advancement of consumer protection. The fourth essay demonstrates that hedonic and normative goal frames, as well as partially gain goal frames, can be regarded as determinants of acceptance. In particular, the findings indicate that hedonic goal frames exert a strong mediating effect, leading to environmentally friendly decisions. This indicates that the acceptance of technologies that integrate digital nudges may be enhanced by addressing the emotional aspect. Furthermore, findings demonstrate that specific forms of nudges can lead to rejection, representing determinants of resistance. With respect to consumers' purchase intentions, social norm nudges have been shown to represent a determinant of resistance.

Essay 5: Decision Aids for Online Size Selection: Insights from Experimental Studies in Apparel Retailing

The findings of the fifth essay indicate that several factors of decision aids influence consumers' intentions to utilize such aids. A greater degree of required data disclosure and a higher level of innovativeness have been found to negatively influence the intention to utilize the decision aid. The findings are consistent with previous research regarding rising concerns about privacy in online shopping contexts. These findings support the privacy calculus theory, which posits that the perceived risks of data disclosure outweigh the perceived benefits of using decision aids (Dinev and Hart, 2006b; Laufer and Wolfe, 1977). In this context, the perceived risk of privacy violations associated with the disclosure of personal data aligns with the findings of prior research (Dinev and Hart, 2006a). Further research could examine more detailed aspects of this theory by evaluating the balance of benefits and risks over time or under different conditions. Longitudinal studies could assess how familiarity with decision aids might reduce resistance. Moreover, the findings of the fifth essay align with those of previous literature by Venkatesh et al. (2003) and Tornatzky and Klein (1982), which similarly demonstrated a negative impact of perceived complexity on the adoption of new technologies. Future researchers might examine how complexity can be reduced in this context and more general in the context of new technologies. Essay five offers support for the basic assumption of IRT (Ram and Sheth, 1989), which posits that a technology's level of innovativeness is inversely correlated with the intention to use it. The literature posits that this reaction might result from the discrepancy between innovative technologies and current consumer habits and values (Kleijnen et al., 2009). The results of the essay indicate that the acceptance and resistance of new technologies are influenced by the perceived cognitive demands and data protection, with consumers balancing the advantages and disadvantages of usage (Dinev and Hart, 2006b). Essay five shows that a high perceived data security when utilizing innovative approaches, and thus a low risk in this regard, seems to act as a determinant of acceptance. Low complexity also represents a determinant of acceptance, indicating that simple and user-friendly systems are preferred. High perceived complexity represents a determinant of resistance due to the high cognitive effort and complex processes involved. A particularly high level of innovativeness and thus a strong deviation from the status quo can also be identified as a determinant of resistance.

Essay 6: Sustainability Assessment of Last-Mile Electrification: A Qualitative Study in Germany

The sixth essay examines the potential of electromobility as an innovative approach to promote sustainability, with a particular focus on the differentiation of ecological, economic, and social sustainability in accordance with the

TBL approach (Elkington, 1994). Results demonstrate that economic considerations seem to shape the stance of last-mile companies and subcontractors towards electric mobility. These companies tend to evaluate the economic sustainability of electric vehicles negatively, citing high acquisition costs and a large-scale changeover as the primary obstacles. In contrast, both ecological and social sustainability are rated more positively or neutrally, yet they tend to play a subordinate role in managers' final decisions. The results demonstrate the necessity of considering additional influencing factors beyond the three pillars of the TBL model. In the context of essay six, these are primarily technical process-related aspects that could also be considered in further theoretical models in the context of sustainable electrification. This finding is also consistent with the results of Jiang et al. (2023). Further research may verify the differentiation into these dimensions and investigate additional dimensions. The research conducted in this dissertation has demonstrated the value of a multidimensional view of sustainability, particularly through the three pillars of the TBL model: ecological, economic, and social sustainability (Elkington, 1997; Purvis et al., 2019). The sixth essay reveals different attitudes towards last-mile electrification among courier, express, and parcel service providers, subcontractors, and drivers. Although all groups identify similar challenges, such as high initial costs and restricted range, there are discrepancies regarding their priorities and experiences. Suppliers primarily focus on economic factors, also affecting the decisions of courier, express, and parcel service providers as well as drivers. However, courier, express, and parcel service providers and drivers also prioritize environmental benefits and the enhanced driving experience. To promote a uniform understanding of the characteristics of electric vehicles, future research might focus on the technical and economic competitiveness of these vehicles by gathering numeric data from real-life experiments. Furthermore, research in this field could concentrate on air quality or noise in urban areas, with the aim of providing evidence of the improvements for environmental and social sustainability. It would be beneficial to include further stakeholders in this context like political and regulatory bodies. Regarding the intention to use electric vehicles, the essay reveals a number of factors that can be identified as determinants of acceptance and resistance, and which can be applied more generally to the use of technologies to promote sustainability. Accordingly, the positive environmental impact and ease of use are demonstrated to be determinants of acceptance. High acquisition costs and uncertainties regarding the range and technical specifications in general are identified as determinants of resistance. Acceptance and resistance are influenced by economic, technical, ecological, and social factors, and the various determinants might be specifically addressed at different levels in future research.

Reference to the General Framework of the Dissertation

Several implications for the general framework of the dissertation can be derived from the six essays presented above. The dissertation offers general insights into the six innovative technologies and approaches of AI, automated purchasing processes, cryptocurrencies, digital nudges, decision aids, and electromobility as also listed in Fig.,4.1. The essays provide both general and more detailed insights into these technologies, with some essays offering more holistic overviews and other essays investigating specific characteristics of the technologies. Results demonstrate the relevance of the considered technologies for businesses and daily life, highlighting the importance of further research and practical application in these contexts. The general framework of the dissertation refers to the UTAUT (Venkatesh et al., 2003) and the IRT (Ram and Sheth, 1989) and their basic assumptions that determinants of acceptance resistance influence the adoption of innovative technologies. Accordingly, the goal of this dissertation was the investigation of the intentions of consumers and companies to use these technologies and to consider various determinants of acceptance and resistance. The determinants identified in the essays were presented above in this chapter and are listed in Fig. 4.1, representing a wide overview about the influencing factors of technology adoption. Future research should further investigate these determinants in the context of innovative technologies, relate them to each other, and consider them by also referring to the interrelation with the variables of the UTAUT. With reference to the IRT, the essays of the dissertation mainly investigate active resistance to innovations (Talke and Heidenreich, 2014) and hereby mainly focus on functional barriers of usage, value, and risk (Talwar et al., 2020). Future researchers should also consider the detected determinants in relation to each other and include the investigation of active resistance as well as psychological barriers in more detail. Fig. 4.1 illustrates an overview about the results of the six essays with regard to the general framework of the dissertation.

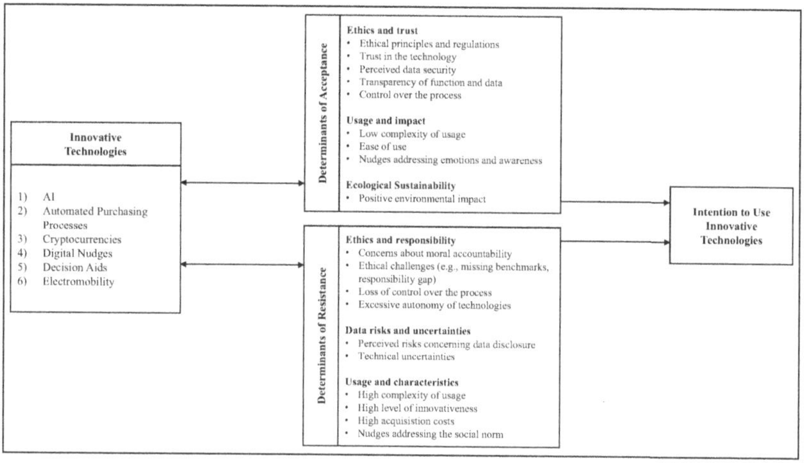

Fig. 4.1 Implications for the General Framework of the Dissertation

4.2 Managerial Implications

Several managerial implications can be derived from the six essays of the dissertation. These implications might be valuable to a wide range of practitioners, including developers, managers of companies interested in using innovative technologies, marketers of retail companies, or politicians. The managerial implications refer to different aspects such as psychological factors influencing consumer behavior as well as determinants of acceptance and resistance of technologies. In particular, the managerial implications are linked to the technologies investigated in this dissertation, being 1) AI; 2) automated purchasing processes; 3) cryptocurrencies; 4) digital nudging; 5) decision aids; and 6) electromobility.

Essay 1: Ethics and Morality in AI—A Systematic Literature Review and Future Research

The systematization of the literature on ethics and morality in the context of AI offers managers an overview about the different topics and aspects discussed in research and the possibility to identify relevant content in this context: ethical and moral implications of AI, approaches of ethical handling of AI, demand for governance and regulations, AI programming requirements, human moral judgment and moral responsibility, moral status of AI and importance of law and politics. Furthermore, the first essay offers a suitable overview of the regulations and recommendations of ethics in AI discussed in literature, being essential for companies working with AI-based technologies or even developing AI. Managers might define and communicate their individual ethical principles and implement these approaches into the usage of AI to support the acceptance of AI with transparency and clarification. Moreover, general as well as specific frameworks should be evolved in this context. This includes general frameworks for an ethical handling of AI, based on human rights and being suitable for different companies, countries, governments, and even cultures. This also includes the development of industry-specific principles and governance to meet the expectations of users of AI as well as the industry-specific requirements. Politicians should provide education programs for companies using innovative technologies like AI to promote an ethical handling of these approaches. Finally, companies and governments using, developing, or implementing AI should be aware of the importance of ethics in general, representing a major determinant of acceptance in this context.

Essay 2: Aspects Driving Customers' Intention to Use Automated Purchasing Processes

The second essay offers managerial implications regarding the use of automated purchasing processes. Marketers should focus on enhancing the trust of consumers into the technologies or new processes implemented. The investigation of trust in essay two concentrates on trust into the process, while trust into the retailer might also be relevant in this context, offering opportunities for future research. Based on this, managers and developers working on autonomous applications should maintain and communicate transparency regarding the handling of data and functions to create trustworthiness for consumers and build a trustworthy relationship (Javaria et al., 2020). Insecure consumers can be convinced to use these autonomous processes by success stories from other customers as positive examples to build trust. In the context of autonomous processes and technologies, keeping control is an important aspect for consumers. Accordingly, companies should develop user-friendly and intuitive interfaces for users and enable customizable settings in addition to transparent communication. For example, consumers preferences and relevant functions should be adjustable manually. Perceived control might be enhanced by including feedback loops, enabling manual interventions in the process, and including personalized preferences. Retailers should start with including partly automated processes and initially not use fully autonomous technologies and processes. In total, retailers should optimize the customer experience by offering transparency regarding new technologies, intuitive operability, and extended opportunities for manual control for consumers about the processes. Furthermore, when trying to offer innovative or autonomous technologies and products, marketers should specifically address consumers seeking for these new products and having lower usage barriers than others (Heidenreich and Handrich, 2015a). Therefore, the target group should be set and marketing campaigns aimed at these groups.

Essay 3: Transferring Trust, Risk and Security on Users' Intention to Trade Cryptocurrency, Using PayPal as an Example

According to the results of the third essay and the investigation of cryptocurrencies, trust and trust-related aspects in existing platforms and technologies can be transferred to new applications. These trust-related aspects are perceived risk and perceived security. When choosing the platform through which new applications will be consumed or distributed, managers should focus on using established platforms that enjoy a high level of consumer trust, low perceived risk, and high perceived security (Casalo et al., 2007). Platform operators should transparently communicate their connection to new applications to reduce uncertainties regarding them, as new technologies can benefit from the positive perception of existing platforms. The context of PayPal as a platform and cryptocurrencies as a new application can also be applied to companies with online shops. Trust and trust-related determinants seem to be transferred from payment platforms to digital currencies. In general, retailers should focus on transparency and the strengthening of security features and therefore might include certificates as well as audits and highlight these aspects in their communication with customers. Furthermore, the focus of retailers to strengthen trust and trust-related aspects should lay on their superior platforms and on linking new technologies with these platforms. Financial service providers and managers included in the distribution of cryptocurrencies should provide educational offers for explaining blockchain and digital currencies, since the knowledge and comfort regarding cryptocurrencies is still low.

Essay 4: Digital Nudging to Reduce Product Returns in E-Commerce: Investigating Consumers' Intentions and Environmental Implications

The dissertation provides insights for managers, marketers, developers, and other decision-makers in the fashion retail industry. Digital nudges can be used for minimizing potential pre-purchase returns and for designing online fashion shopping more sustainable. Marketers can use various digital nudges such as default, incentive, and framing to influence consumers' decision making (Bucher et al., 2016). Especially framing supports pro-environmental consumer behavior and should be considered by marketers and designers for not only influencing consumers' decision-making, but also shaping the consumer experience more positively and intuitively (Weinmann et al., 2016). It is important for marketers to maintain consumer autonomy and strike a balance between digital nudges and independent consumer decisions. Therefore, ethical standards should be considered when implementing digital nudges. Retailers should rely on ethical regulations and principles for the use of digital nudges and politicians should implement further regulations for protecting consumers' autonomy. Overall, digital nudges can serve as an important starting point for the sustainable design of online commerce. Using digital nudges with ecological information can shape consumers' awareness about sustainable behavior in online shopping in general. In addition to the environmental benefits of reduced returns, companies benefit economically from implementing digital nudges. Applying digital nudges to other contexts, policymakers could utilize them to encourage pro-environmental behavior in various contexts like mobility or regional purchasing.

Essay 5: Decision Aids for Online Size Selection: Insights from Experimental Studies in Apparel Retailing

Essay five shows that marketers and retailers should be aware of several factors when implementing decision aids in online retailing. Consumers tend to reject decision aids that require a high level of data input and consumers reject sharing their personal data with these aids. To ensure transparency, marketers should inform consumers about the usage, storage, and security of their data by technologies to minimize the concerns about data disclosure. Furthermore, the amount of personal data input required to use technologies should be as low as possible. Retailers may also implement technologies that do not change too much from the known status-quo and therefore are not rejected by consumers due to their innovativeness. Managers should also be aware of the complexity of technologies they implement. Consumers might negatively evaluate high complexity of new approaches. Technologies should be designed and communicated in a way being clear, concise, and objective. To promote the adoption of decision-making aids and similar technologies, retail companies are recommended to provide clear instructions on their proper use (Tan et al., 2012). Results of essay five also illustrate that higher innovativeness of these aids only affects consumers' intention to use these aids, but not to transact in the online shop in general. Therefore, retailers may offer the opportunity of using decision aids and recommendation agents but make them mandatory for online retailing.

Essay 6: Sustainability Assessment of Last-Mile Electrification: A Qualitative Study in Germany

The sixth essay of this dissertation demonstrates the importance for companies to involve stakeholders at different levels in decision-making processes regarding the use of new technologies and approaches like the switch to electromobility in the logistics sector. It is important for company managers to draw on the experience of active users, and decisions should not be made solely at one hierarchical level. Similarly, politicians should consider the perspectives of all stakeholders when making legal decisions. When introducing new innovative approaches, it is important for company managers, associations, and politicians to consider the different user levels. Communication should not only focus on ecological sustainability, but also on economic and social sustainability as well as technical aspects and corresponding benefits when switching to new technologies. Considering the total cost of ownership, electromobility is already cost-competitive (Tsakalidis et al., 2020) and provides sufficient range for the logistics sector (Iwan et al., 2021) according to literature. Therefore, communication regarding the transition to electromobility should include aspects of technical and economic sustainability highlighting the competitiveness of electric vehicles. Company managers need to be willing to trust these facts and not allow their attitudes being influenced by rooted patterns of thinking. Politicians, associations, and electric vehicles manufacturers can all support communication efforts. In general, providers of new innovations aimed at increasing ecological sustainability should consider economic and social sustainability and incorporate them into their communication strategies. To minimize uncertainty regarding innovative technologies, test editions might be offered, and the respective technologies should be made available for interested companies and consumers.

4.3 Concluding Remarks

Based on the general research framework, the dissertation contributes to the literature on intentions of consumers and partly companies of using the following innovative technologies: AI, automated shopping processes, cryptocurrencies, digital nudges, decision aids, and electromobility. The findings of the six essays are based on a systematic literature review, quantitative research studies, and interviews. Different research methods and samples were used in the studies to offer an as good as holistic overview about the context. The dissertation relates to the academic literature of marketing and retailing as well as information systems and offers implications for practitioners mainly working in a marketing or technology context. Referring to the Unified Theory of Acceptance and Use of Technology and the Innovation Resistance Theory, the dissertation highlights the importance of various determinants of acceptance and resistance for the adoption of innovative technologies. Based on the results, an ethical framework and the focus on sustainability, among others, represent significant determinants for a successful adoption of new technologies. Some limitations should be mentioned for a better classification of the results of the dissertation. While each essay contains own limitations, only the overarching limitations are summarized. The technologies investigated in the essays were considered in theoretical experiments and were not actively used by the participants. Furthermore, the data of the empirical studies of the essays was solely collected in Germany and mostly with a local focus, requiring the investigation in larger areas and other countries in the future. The different determinants of acceptance and resistance investigated in the dissertation were chosen with reference to previous literature and practical relevance. Accordingly, the choice of determinants is not concluding and should be extended in future research. Overall and in conclusion, the dissertation represents a comprehensive investigation of the intention to use different innovative technologies, and hereby the in-depth consideration of various determinants of acceptance and resistance.

References

Abdulla, G.M., Borar, S., 2017. Size Recommendation System for Fashion E-commerce. KDD Workshop on Machine Learning Meets Fashion.

About Amazon Team, 2020. Our biggest commitment to electromobility in Germany to date: The last mile is green in Essen. Amazon <https://www.aboutamazon.eu/news/sustainability/our-biggest-commitment-to-electromobility-in-germany-to-date-the-last-mile-is-green-in-essen> (accessed 20 October 2023).

Abràmoff, M.D., Tobey, D., Char, D.S., 2020. Lessons Learned About Autonomous AI: Finding a Safe, Efficacious, and Ethical Path Through the Development Process. American Journal of Ophthalmology 214, 134–142.

Abrams, M., Abrams, J., Cullen, P., Goldstein, L., 2019. Artificial Intelligence, Ethics, and Enhanced Data Stewardship. IEEE Secur. Privacy 17 (2), 17–30.

acea, 2023. New commercial vehicle registrations: vans +11.2%, trucks +20%, buses + 15% in the first half of 2023 <https://www.acea.auto/cv-registrations/new-commercial-vehicle-registrations-vans-11-6-trucks-20-buses-15-in-the-first-half-of-2023/> (accessed 19 October 2023).

Adamson, G., Havens, J.C., Chatila, R., 2019. Designing a Value-Driven Future for Ethical Autonomous and Intelligent Systems. Proceedings of the IEEE 107 (3), 518–525.

Aizenberg, E., van den Hoven, J., 2020. Designing for human rights in AI. Big Data & Society 7 (2), 205395172094956.

Ajzen, I., 1985. From Intentions to Actions: A Theory of Planned Behavior, in: Kuhl, J., Beckmann, J. (Eds.), Action Control. From Cognition to Behavior. Springer, Berlin, Heidelberg, New York, Tokyo, 11–39.

Ajzen, I., 1991. The theory of planned behavior. Organizational Behavior and Human Decision Processes 50 (2), 179–211.

Ajzen, I., Fishbein, M., 1975. Predicting and understanding consumer behavior: Attitude-behavior correspondence, in: Ajzen, I., Fishbein, M. (Eds.), Understanding Attitudes and Predicting Social Behavior. Englewood Cliffs, NJ, Prentice Hall, 148–172.

Allcott, H., Kessler, J.B., 2019. The Welfare Effects of Nudges: A Case Study of Energy Use Social Comparisons. American Economic Journal: Applied Economics 11 (1), 236–276.

Allcott, H., Taubinsky, D., 2015. Evaluating Behaviorally Motivated Policy: Experimental Evidence from the Lightbulb Market. American Economic Review 105 (8), 2501–2538.

J. Selter, *Acceptance and Resistance of Innovative Technologies*, Handel und Internationales Marketing Retailing and International Marketing,
https://doi.org/10.1007/978-3-658-48594-8

Almeida, P.G.R. de, dos Santos, C.D., Farias, J.S., 2021. Artificial Intelligence Regulation: a framework for governance. Ethics and Information Technology, 1–21.

Anic, I.-D., Škare, V., Kursan Milaković, I., 2019. The determinants and effects of online privacy concerns in the context of e-commerce. Electronic Commerce Research and Applications 36, 100868.

Anosike, A., Loomes, H., Udokporo, C.K., Garza-Reyes, J.A., 2023. Exploring the challenges of electric vehicle adoption in final mile parcel delivery. International Journal of Logistics Research and Applications 26, 683–707.

Anshari, M., Almunawar, M.N., Masri, M., Hrdy, M., 2021. Financial Technology with AI-Enabled and Ethical Challenges. Soc 58 (3), 189–195.

Antao, D., 2019. Technological Advancement, Artificial Intelligence and the Relevance of Classroom Teachers. IJRAR-International Journal of Research and Analytical Reviews (IJRAR) 6 (2), 353–361.

Anthis, J.R., Paez, E., 2021. Moral circle expansion: A promising strategy to impact the far future. Futures 130, 102756.

Antle, A.N., Kitson, A., 2021. 1,2,3,4 tell me how to grow more: A position paper on children, design ethics and biowearables. International Journal of Child-Computer Interaction 30, 100328.

Ardani, E.G., Juliadi, R., 2019. The Threat of Personal Information Trading in the Online Network and Information Technology. International Journal of Multicultural and Multireligious Understanding 6 (9), 49–55.

Arias-Oliva, M., Pelegrín-Borondo, J., Matías-Clavero, G., 2019. Variables Influencing Cryptocurrency Use: A Technology Acceptance Model in Spain. Frontiers in Psychology 10, 475.

Arifin, A.A., Lennerfors, T.T., 2022. Ethical aspects of voice assistants: a critical discourse analysis of Indonesian media texts. Journal of Information, Communication and Ethics in Society 20 (1), 18–36.

Arli, D., van Esch, P., Bakpayev, M., Laurence, A., 2021. Do consumers really trust cryptocurrencies? Marketing Intelligence & Planning 39 (1), 74–90.

Armitage, C.J., Conner, M., 2001. Efficacy of the theory of planned behaviour: A meta-analytic review. British Journal of Social Psychology 40 (4), 471–499.

Arnold, T., Scheutz, M., 2018. The "big red button" is too late: an alternative model for the ethical evaluation of AI systems. Ethics Inf Technol (Ethics and Information Technology 20 (1), 59–69.

Arvan, M., 2018. Mental time-travel, semantic flexibility, and A.I. ethics. AI & Soc, 1–20.

Asaro, P.M., 2019. AI Ethics in Predictive Policing: From Models of Threat to an Ethics of Care. IEEE Technol. Soc. Mag. 38 (2), 40–53.

Asdecker, B., Karl, D., 2022. Shedding Some Light on the Reverse Part of E-Commerce: A Systematic Look Into the Black Box of Consumer Returns in Germany. European Journal of Management 22, 59–81.

Asdecker, B., Karl, D., Sucky, E., 2016. Retourenmanagement als Erfolgsfaktor des Handels, in: Schallmo, D., Rusnjak, A., Anzengruber, J., Werani, T., Jünger, M. (Eds.), Digitale Transformation von Geschäftsmodellen. Grundlagen, Instrumente und Best Practices. Springer Gabler, Wiesbaden, Wiesbaden, 595–612.

Ashok, M., Madan, R., Joha, A., Sivarajah, U., 2022. Ethical framework for Artificial Intelligence and Digital technologies. International Journal of Information Management 62, 102433.

ASOS, 2024. What is Fit Assistant? <https://www.asos.com/customer-care/product-stock/what-is-fit-assistant/> (accessed 1 July 2024).

Attard-Frost, B., los Ríos, A. de, Walters, D.R., 2023. The ethics of AI business practices: a review of 47 AI ethics guidelines. AI and Ethics 3 (2), 389–406.

Autili, M., Di Ruscio, D., Inverardi, P., Pelliccione, P., Tivoli, M., 2019. A Software Exoskeleton to Protect and Support Citizen's Ethics and Privacy in the Digital World. IEEE Access 7, 62011–62021.

Awad, E., Dsouza, S., Kim, R., Schulz, J., Henrich, J., Shariff, A., Bonnefon, J.-F., Rahwan, I., 2018. The Moral Machine experiment. Nature 563 (7729), 59–64.

Awad, N.F., Krishnan, M.S., 2006. The Personalization Privacy Paradox: An Empirical Evaluation of Information Transparency and the Willingness to Be Profiled Online for Personalization. MIS Quarterly 30 (1), 13–28.

Bærøe, K., Miyata-Sturm, A., Henden, E., 2020. How to achieve trustworthy artificial intelligence for health. Bulletin of the World Health Organization 98 (4), 257–262.

Baier, L., Rackow, T., Donhauser, T., Pfeffer, D., Schuderer, P., Franke, J., 2016. Logistical Integration of Smart Homes for Automated Consumer Goods Supply Based on Smart Refrigerators. AEF 19, 107–115.

Balaji, M.S., Roy, S.K., 2017. Value co-creation with Internet of things technology in the retail industry. Journal of Marketing Management 33 (1–2), 7–31.

Baltes, M.M., Hayward, S.C., 1976. Application and evaluation of strategies to reduce pollution: Behavioral control of littering in a football stadium. Journal of Applied Psychology 60 (4), 501–506.

Bamberg, S., Möser, G., 2007. Twenty years after Hines, Hungerford, and Tomera: A new meta-analysis of psycho-social determinants of pro-environmental behaviour. Journal of Environmental Psychology 27 (1), 14–25.

Bankins, S., 2021. The ethical use of artificial intelligence in human resource management: a decision-making framework. Ethics and Information Technology 23 (4), 841–854.

Banks, J., 2020. Good Robots, Bad Robots: Morally Valenced Behavior Effects on Perceived Mind, Morality, and Trust. Int J of Soc Robotics, 1–18.

Bansal, H.S., Irving, P.G., Taylor, S.F., 2004. A Three-Component Model of Customer to Service Providers. Journal of the Academy of Marketing Science 32 (3), 234–250.

Baron, R.M., Kenny, D.A., 1986. The moderator-mediator variable distinction in social psychological research: Conceptual, strategic, and statistical considerations. Journal of Personality and Social Psychology 51 (6), 1173.

Bartneck, C., Lütge, C., Wagner, A., Welsh, S., 2019. Ethik in KI und Robotik. Carl Hanser Verlag GmbH Co KG, München.

Bauer, W.A., 2020. Virtuous vs. utilitarian artificial moral agents. AI & Soc 35 (1), 263–271.

Baum, S.D., 2020. Social choice ethics in artificial intelligence. AI & Soc 35 (1), 165–176.

Bayer, S., Gimpel, H., Rau, D., 2021. IoT-commerce—opportunities for customers through an affordance lens. Electron Markets 31 (1), 27–50.

BdKEP, 2023. Interessenvertretung, Innovationen und Einkaufsvorteile für die KEP Branche <https://bdkep.de/> (accessed 18 October 2023).

Beauchamp, T.L., 2003. The Nature of Applied Ethics, in: Frey, R.G., Wellman, C.H. (Eds.), A Companion to Applied Ethics. John Wiley & Sons, New York, NY, 1–16.

Behrmann, E., 2019. A Dead End for Fossil Fuel in Europe's City Centers <https://www.bloomberg.com/news/articles/2019-07-26/a-dead-end-for-fossil-fuel-in-europe-s-city-centers?in_source=embedded-checkout-banner> (accessed 5 September 2023).

Belanche, D., Casaló, L.V., Flavián, C., Schepers, J., 2014. Trust transfer in the continued usage of public e-services. Information & Management 51 (6), 627–640.

Belk, R., 2021. Ethical issues in service robotics and artificial intelligence. The Service Industries Journal 41 (13–14), 860–876.

Bellis, E. de, Johar, G.V., 2020. Autonomous Shopping Systems: Identifying and Overcoming Barriers to Consumer Adoption. Journal of Retailing 96 (1), 74–87.

Bentham, J., 1948 (orig. 1789). Introduction to the Principles of Morals and Legislation. Oxford University Press.

Benzmüller, C., Parent, X., van der Torre, L., 2020. Designing normative theories for ethical and legal reasoning: LogiKEy framework, methodology, and tool support. Artificial Intelligence 287, 103348.

Berger, J., Heath, C., 2007. Where Consumers Diverge from Others: Identity Signaling and Product Domains. Journal of Consumer Research 34 (2), 121–134.

Berger, M., Greinacher, E., Wolf, L., 2022. Digital Nudging to Promote Energy Conservation Behavior—Framing and Default Rules in a Smart Home App. Proceedings of the 2022 European Conference on Information Systems, Timisoara.

Berger, M., Müller, C., Nüske, N., 2020. Digital Nudging in Online Grocery Stores-Towards Ecologically Sustainable Nutrition. Proceedings of the 2020 International Conference on Information Systems, Hyderabad, India.

Bernecker, T., Lesemann, L., Heinzelmann, J., Hujer, J., Hartwig, M., Pfeifer, F., Wendorf, A., Kamenz, S., 2022. Nachhaltige Logistik der Kurier-, Express- und Paketdienste (KEP). e-mobil BW GmbH, Stuttgart.

Bhatnagar, A., Ghose, S., 2004. Segmenting consumers based on the benefits and risks of Internet shopping. Journal of Business Research 57 (12), 1352–1360.

BIEK, 2023. Bundesverband Paket & Express Logistik <https://www.biek.de/startseite.html> (accessed 18 October 2023).

Blut, M., 2016. E-Service Quality: Development of a Hierarchical Model. Journal of Retailing 92 (4), 500–517.

Blut, M., Ghiassaleh, A., Wang, C., 2023. Testing the performance of online recommendation agents: A meta-analysis. Journal of Retailing 99 (3), 440–459.

Bock, A., 2020. Umfrage: Immer weniger Deutsche nutzen Kryptowährungen <https://www.bearingpoint.com/de-de/ueber-uns/pressemitteilungen-und-medienberichte/pressemitteilungen/umfrage-immer-weniger-deutsche-nutzen-kryptowaehrungen/> (accessed 21 August 2023).

Boeuf, B., Sénécal, S., 2013. Online international outshopping experience: Proposition of a research model. Recherche et Applications en Marketing 28 (3), 110–119.

Bogosian, K., 2017. Implementation of Moral Uncertainty in Intelligent Machines. Minds & Machines 27 (4), 591–608.

Bohr, J., Bashir, M., 2014. Who Uses Bitcoin? An exploration of the Bitcoin community. Twelfth Annual International Conference on Privacy, Security and Trust in Toronto, Canada.

Boradbent, E., 2017. Interactions with Robots: The Truths we Reveal about Ourselves. Annual Review of Psychology 68 (1), 627–652.

Borau, S., Otterbring, T., Laporte, S., Fosso Wamba, S., 2021. The most human bot: Female gendering increases humanness perceptions of bots and acceptance of AI. Psychology & Marketing 38 (7), 1052–1068.

Borenstein, J., Herkert, J.R., Miller, K.W., 2019. Self-Driving Cars and Engineering Ethics: The Need for a System Level Analysis. Sci Eng Ethics 25 (2), 383–398.

Bosona, T., 2020. Urban freight last mile logistics—Challenges and opportunities to improve sustainability: A literature review. Sustainability 12 (21), 8769.

Botti, S., Iyengar, S.S., 2006. The Dark Side of Choice: When Choice Impairs Social Welfare. Journal of Public Policy & Marketing 25 (1), 24–38.

Botti, S., McGill, A.L., 2011. The Locus of Choice: Personal Causality and Satisfaction with Hedonic and Utilitarian Decisions. Journal of Consumer Research 37 (6), 1065–1078.

Bower, A.B., Maxham, J.G., 2012. Return Shipping Policies of Online Retailers: Normative Assumptions and the Long-Term Consequences of Fee and Free Returns. Journal of Marketing 76 (5), 110–124.

Bozzi, C., Neves, M., Alvao, C.M., 2022. Fashio E-Tail and the Impact of Returns: Mapping Processes and the Consumer Journey towards More Sustainable Practices. Sustainability 14 (9), 5328.

Brady, A.P., Neri, E., 2020. Artificial Intelligence in Radiology-Ethical Considerations. Diagnostics (Basel, Switzerland) 10 (4), 231.

Braun, H., 2019. Eckpunkte für das Klimaschutzprogramm 2030. Bundesregierung.

Brendel, A.B., Mirbabaie, M., Lembcke, T.-B., Hofeditz, L., 2021. Ethical Management of Artificial Intelligence. Sustainability 13 (4), 1974.

Brennan, S., Bravo Gonzalez, R., Puznlak-Holford, M., Kilsby, R., Subramoni, A., 2023. EU regulations drives the sustainability transition: Sustainability regulation outlook <https://www2.deloitte.com/us/en/insights/environmental-social-governance/transition-to-net-zero.html> (accessed 5 February 2024).

Brezo, F., Bringas, P.G., 2012. Issues and risks associated with cryptocurrencies such as Bitcoin. The Second International Conference on Social Eco-Informatics, 20–26.

Briner, R.B., Denyer, D., 2012. Systematic Review and Evidence Synthesis as a Practice and Scholarship Tool. Oxford University Press.

Brown, B., 2001. Studying the internet experience. HP laboratories technical report HPL 49.

Brown, J.E., Halpern, J., 2021. AI chatbots cannot replace human interactions in the pursuit of more inclusive mental healthcare. SSM—Mental Health 1, 100017.

Browne, R., 2021. PayPal launches its cryptocurrency service in the UK <https://www.cnbc.com/2021/08/22/paypal-crypto-service-launches-in-the-uk.html> (accessed 21 August 2023).

Brożek, B., Janik, B., 2019. Can artificial intelligences be moral agents? New Ideas in Psychology 54, 101–106.

Bruzzone, F., Cavallaro, F., Nocera, S., 2021. The integration of passenger and freight transport for first-last mile operations. Transport Policy 100, 31–48.

Bryson, J.J., 2018. Patiency is not a virtue: the design of intelligent systems and systems of ethics. Ethics Inf Technol 20 (1), 15–26.

Bucher, T., Collins, C., Rollo, M.E., McCaffrey, T.A., Vlieger, N. de, van der Bend, D., Truby, H., Perez-Cueto, F.J.A., 2016. Nudging consumers towards healthier choices: a systematic review of positional influences on food choice. British Journal of Nutrition 115 (12), 2252–2263.

Burton, E., Goldsmith, J., Koenig, S., Kuipers, B., Mattei, N., Walsh, T., 2017. Ethical Considerations in Artificial Intelligence Courses. AIMag 38 (2), 22–34.

Burton, S., Garretson, J.A., Velliquette, A.M., 1999. Implications of Accurate Usage of Nutrition Facts Panel Information for Food Product Evaluations and Purchase Intentions. Journal of the Academy of Marketing Science 27 (4), 470–480.

Burton, S., Habli, I., Lawton, T., McDermid, J., Morgan, P., Porter, Z., 2020. Mind the gaps: Assuring the safety of autonomous systems from an engineering, ethical, and legal perspective. Artificial Intelligence 279, 103201.

Buruk, B., Ekmekci, P.E., Arda, B., 2020. A critical perspective on guidelines for responsible and trustworthy artificial intelligence. Med Health Care and Philos 23 (3), 387–399.

BVL, transport logistic, 2019. Männerdomäne?! Chancen für und mit Frauen in der Logistik. Bundesvereinigung Logistik e. V; Messe München GmbH transport logistic <https://www.bvl.de/files/1951/1988/2128/Begleitende_Publikation_zur_Session_Maennerdomaene.pdf> (accessed 12 July 2023).

Calvo, P., 2020. The ethics of Smart City (EoSC): moral implications of hyperconnectivity, algorithmization and the datafication of urban digital society. Ethics Inf Technol 22 (2), 141–149.

Campbell, C., Sands, S., Ferraro, C., Tsao, Hsiu-Yuan (Jody), Mavrommatis, A., 2020. From data to action: How marketers can leverage AI. Business Horizons 63 (2), 227–243.

Cao, X., Yu, L., Liu, Z., Gong, M., Adeel, L., 2018. Understanding mobile payment users' continuance intention: a trust transfer perspective. Internet Research 28 (2), 456–476.

Capgemini Research Institute, 2023. A World in Balance 2023: Heightened Sustainability Awareness Yet Lagging Actions.

Caraban, A., Karapanos, E., Gonçalves, D., Campos, P., 2019. 23 Ways to Nudge: A Review of Technology-Mediated Nudging in Human-Computer Interaction. Proceedings of the 2019 CHI Conference on Human Factors in Computing Systems, New York, USA.

Carillo, R.M., 2020. Artificial intelligence: From ethics to law. Telecommunications Policy 44 (6), 101937.

Casalo, L.V., Flavian, C., Guinaliu, M., 2007. The role of security, privacy, usability and reputation in the development of online banking. Online Information Review 31 (5), 583–603.

Cath, C., Wachter, S., Mittelstadt, B., Taddeo, M., Floridi, L., 2018. Artificial Intelligence and the 'Good Society': the US, EU, and UK approach. Sci Eng Ethics 24 (2), 505–528.

Cervantes, J.-A., López, S., Rodríguez, L.-F., Cervantes, S., Cervantes, F., Ramos, F., 2020a. Artificial Moral Agents: A Survey of the Current Status. Sci Eng Ethics 26 (2), 501–532.

Cervantes, S., López, S., Cervantes, J.-A., 2020b. Toward ethical cognitive architectures for the development of artificial moral agents. Cognitive systems research 64, 117–125.

Chang, S.-C., Tung, F.-C., 2008. An empirical investigation of students' behavioural intentions to use the online learning course websites. British Journal of Educational Technology 39 (1), 71–83.

Chaudhuri, A., Holbrook, M.B., 2002. Product-class effects on brand commitment and brand outcomes: The role of brand trust and brand affect. Journal of Brand Management 10 (1), 33–58.

Chauhan, C., Gullapalli, R.R., 2021. Ethics of AI in Pathology: Current Paradigms and Emerging Issues. The American journal of pathology 191 (10), 1673–1683.

Cheah, E.-T., Fry, J., 2015. Speculative bubbles in Bitcoin markets? An empirical investigation into the fundamental value of Bitcoin. Economics Letters 130, 32–36.

Chellappa, R.K., Pavlou, P.A., 2002. Perceived information security, financial liability and consumer trust in electronic commerce transactions. Logistics Information Management 15 (5/6), 358–368.

Chen, C., 2006. Identifying Significant Factors Influencing Consumer Trust in an Online Travel Site. Cognizant Communication Corporation.

Chen, C., Xu, X., Frey, S., 2016. Who wants solar water heaters and alternative fuel vehicles? Assessing social–psychological predictors of adoption intention and policy support in China. Energy Research & Social Science 15, 1–11.

Cheng, T.E., Lam, D.Y.C., Yeung, A.C., 2006. Adoption of internet banking: an empirical study in Hong Kong. Decision Support Systems 42 (3), 1558–1572.

Chiu, C.-M., Wang, E.T.G., Fang, Y.-H., Huang, H.-Y., 2014. Understanding customers' repeat purchase intentions in B2C e-commerce: the roles of utilitarian value, hedonic value and perceived risk. Information Systems Journal 24 (1), 85–114.

Chounta, I.-A., Bardone, E., Raudsep, A., Pedaste, M., 2021. Exploring Teachers' Perceptions of Artificial Intelligence as a Tool to Support their Practice in Estonian K-12 Education. Int J Artif Intell Educ, 1–31.

Chui, M., Roberts, R., Yee, L., McKinsey & Company, 2024. The Technology Trends That Executives Should Know: Which developments in technology will matter most for companies in the year ahead? <https://partners.wsj.com/mckinsey/the-technology-trends-that-executives-should-know/> (accessed 1 October 2024).

Ciaian, P., Rajcaniova, M., Kancs, d., 2016. The digital agenda of virtual currencies: Can BitCoin become a global currency? Information Systems and e-Business Management 14 (4), 883–919.

Cialdini, R.B., Reno, R.R., Kallgren, C.A., 1990. A focus theory of normative conduct: Recycling the concept of norms to reduce littering in public places. Journal of Personality and Social Psychology 58 (6), 1015–1026.

Cichocki, A., Kuleshov, A.P., 2021. Future Trends for Human-AI Collaboration: A Comprehensive Taxonomy of AI/AGI Using Multiple Intelligences and Learning Styles. Computational Intelligence and Neuroscience 2021, 1–21.

Clark, J., Essex, A., 2012. CommitCoin: Carbon Dating Commitments with Bitcoin. International conference on financial cryptography and data security, Berlin, Heidelberg, 390–398.

Clarke, R., 2019. Principles and business processes for responsible AI. Computer Law & Security Review 35 (4), 410–422.

Claudy, M.C., Garcia, R., O'Driscoll, A., 2015. Consumer resistance to innovation—a behavioral reasoning perspective. Journal of the Academy of Marketing Science 43 (4), 528–544.

Clemons, E.K., Croson, D.C., Weber, B.W., 1996. Reengineering Money: The Mondex Stored Value Card and Beyond. International Journal of Electronic Commerce 1 (2), 5–31.

Coeckelbergh, M., 2021. Narrative responsibility and artificial intelligence. AI & SOCIETY, 1–14.

Cofinpro AG, 2018. FinWeb Barometer—Digitales Banking 2018—Bevölkerungsrepräsentative Studie unter 2.000 Bundesbürgern <https://www.cofinpro.de/pdf/contentful/MmgKMzN7KoUkqGyEsiU2Y.pdf> (accessed 13 April 2021).

Cohen, J., 1977. Statistical power analysis for the behavioral sciences, New York.

CoinDesk, 2015. New CoinDesk Report Reveals Who Really Uses Bitcoin <https://www.coindesk.com/new-coindesk-report-reveals-who-really-uses-bitcoin> (accessed 21 August 2023).

Collier, J.E., Sherrell, D.L., 2010. Examining the influence of control and convenience in a self-service setting. Journal of the Academy of Marketing Science 38 (4), 490–509.

Colombino, T., Gallo, D., Shreepriya, S., Im, Y., Cha, S., 2021. Ethical Design of a Robot Platform for Disabled Employees: Some Practical Methodological Considerations. Frontiers in Robotics and AI 8, 235.

Coman, A., Aha, D.W., 2018. AI Rebel Agents. AIMag 39 (3), 16–26.

Comi, A., Savchenko, L., 2021. Last-mile delivering: Analysis of environment-friendly transport. Sustainable Cities and Society 74, 103213.

Connolly, T., Zeelenberg, M., 2002. Regret in Decision Making. Current Directions in Psychological Science 11 (6), 212–216.

Constantinides, E., Kahlert, M., Vries, S.A. de, 2017. The Relevance of Technological Autonomy in the Acceptance of IoT Services in Retail. 2nd International Conference on Internet of Things, Data and Cloud Computing, ICC.

Conti, M., Sandeep Kumar, E., Lal, C., Ruj, S., 2018. A Survey on Security and Privacy Issues of Bitcoin. IEEE Communications Surveys & Tutorials 20 (4), 3416–3452.

Corbet, S., Lucey, B., Urquhart, A., Yarovaya, L., 2019. Cryptocurrencies as a financial asset: A systematic analysis. International Review of Financial Analysis 62, 182–199.

Creswell, J.W., 2013. Research Design: Qualitative, quantitative, and mixed methods approaches, 4th ed. sage, Los Angeles, Calif.

Cullinane, S., Browne, M., Karlsson, E., Wang, Y., 2019. Retail Clothing Returns: A Review of Key Issues, in: Wells, P. (Ed.), Contemporary Operations and Logistics. Palgrave Macmillan, Cham, 301–322.

Culnan, M.J., Armstrong, P.K., 1999. Information Privacy Concerns, Procedural Fairness, and Impersonal Trust: An Empirical Investigation. Organization Science 10 (1), 104–115.

Cunneen, M., Mullins, M., Murphy, F., 2019a. Autonomous Vehicles and Embedded Artificial Intelligence: The Challenges of Framing Machine Driving Decisions. Applied Artificial Intelligence 33 (8), 706–731.

Cunneen, M., Mullins, M., Murphy, F., Gaines, S., 2019b. Artificial driving intelligence and moral agency: Examining the decision ontology of unavoidable road traffic accidents through the prism of the trolley dilemma. Applied Artificial Intelligence 33 (3), 267–293.

Cunneen, M., Mullins, M., Murphy, F., Shannon, D., Furxhi, I., Ryan, C., 2020. Autonomous Vehicles and Avoiding the Trolley (Dilemma): Vehicle Perception, Classification, and the Challenges of Framing Decision Ethics. Cybernetics and Systems 51 (1), 59–80.

Cvik, Š., 2021. Categorization and challenges of utilitarianisms in the context of artificial intelligence. AI & Soc, 1–7.

D'Acquisto, G., 2020. On conflicts between ethical and logical principles in artificial intelligence. AI & Soc 35 (4), 895–900.

Dabholkar, P.A., 2006. Factors Influencing Consumer Choice of a "Rating Web Site": An Experimental Investigation of an Online Interactive Decision Aid. Journal of Marketing Theory and Practice 14 (4), 259–273.

Dabholkar, P.A., Bagozzi, R.P., 2002. An Attitudinal Model of Technology-Based Self-Service: Moderating Effects of Consumer Traits and Situational Factors. Journal of the Academy of Marketing Science 30 (3), 184–201.

Dachser, 2018. Emissionsfrei in die City <https://www.dachser.de/de/mediaroom/Emissionsfrei-in-die-City-773> (accessed 22 January 2021).

Dahlberg, T., Mallat, N., Ondrus, J., Zmijewska, A., 2008. Past, present and future of mobile payments research: A literature review. Electronic Commerce Research and Applications 7 (2), 165–181.

Daina, N., Sivakumar, A., Polak, J.W., 2017. Electric vehicle charging choices: Modelling and implications for smart charging services. Transportation Research Part C: Emerging Technologies 81 (Part 1), 36–56.

Danaher, J., 2019. The rise of the robots and the crisis of moral patiency. AI & Soc 34 (1), 129–136.

Daugherty, P.R., Wilson, H.J., 2018. Human + machine: Reimagining work in the age of AI. Harvard Business Press, Brighton.

Davenport, T., Guha, A., Grewal, D., Bressgott, T., 2020. How artificial intelligence will change the future of marketing. Journal of the Academy of Marketing Science 48 (1), 24–42.

Davis, F.D., 1989. Perceived Usefulness, Perceived Ease of Use, and User Acceptance of Information Technology. MIS Quarterly 13 (3), 319.

Deges, F., 2017. Retourenmanagement im Online-Handel. Springer Fachmedien Wiesbaden, Wiesbaden.

Delgado-Ballester, E., Hernández-Espallardo, M., 2008. Effect of Brand Associations on Consumer Reactions to Unknown On-Line Brands. International Journal of Electronic Commerce 12 (3), 81–113.

Delmas, M.A., Lessem, N., 2014. Saving power to conserve your reputation? The effectiveness of private versus public information. Journal of Environmental Economics and Management 67 (3), 353–370.

Deloitte, 2024. The Sustainable Consumer 2023 <https://www2.deloitte.com/uk/en/pages/consumer-business/articles/sustainable-consumer.html> (accessed 4 February 2024).

Dhamodharan, R., 2021. Warum Mastercard Crypto in sein Netzwerk bringt <https://www.mastercard.com/news/europe/de-de/blog/de-de/2020/warum-mastercardcrypto-in-sein-netzwerk-bringt/> (accessed 16 April 2021).

Diaz-Ramirez, J., Zazueta-Nassif, S., Galarza-Tamez, R., Prato-Sanchez, D., Huertas, J.I., 2023. Characterization of urban distribution networks with light electric freight vehicles. Transportation Research Part D: Transport and Environment 119, 103719.

Dinev, Hart, 2006a. Privacy Concerns and Levels of Information Exchange: An Empirical Investigation of Intended e-Services Use. e-Service Journal 4 (3), 25.

Dinev, T., Hart, P., 2006b. An Extended Privacy Calculus Model for E-Commerce Transactions. Information Systems Research 17 (1), 61–80.

Dobbe, R., Krendl Gilbert, T., Mintz, Y., 2021. Hard choices in artificial intelligence. Artificial Intelligence 300, 103555.

Dolan, P., Hallsworth, M., Halpern, D., King, D., Metcalfe, R., Vlaev, I., 2012. Influencing behaviour: The mindspace way. Journal of Economic Psychology 33 (1), 264–277.

Donmez-Turan, A., Kiliclar, I.E., 2021. The analysis of pro-environmental behaviour based on ecological worldviews, environmental training/ knowledge and goal frames. Journal of Cleaner Production 279, 123518.

Du, S., 2021. Reimagining the Future of Technology: "The Social Dilemma" Review. J Bus Ethics, 1–3.

Du, S., Xie, C., 2021. Paradoxes of artificial intelligence in consumer markets: Ethical challenges and opportunities. Journal of Business Research 129, 961–974.

Dubljević, V., 2020. Toward Implementing the ADC Model of Moral Judgment in Autonomous Vehicles. Sci Eng Ethics 26 (5), 2461–2472.

Dur, R., Non, A., Roelfsema, H., 2010. Reciprocity and incentive pay in the workplace. Journal of Economic Psychology 31 (4), 676–686.

Dwivedi, Y.K., Hughes, L., Ismagilova, E., Aarts, G., Coombs, C., Crick, T., Duan, Y., Dwivedi, R., Edwards, J.S., Eirug, A., Galanos, V., Ilavarasan, P.V., Janssen, M., Jones, P., Kumar Kar, A., Kizgin, H., Kronemann, B., Lal, B., Lucini, B., Medaglia, R., Williams Michael D., 2021. Artificial Intelligence (AI): Multidisciplinary perspectives on emerging challenges, opportunities, and agenda for research, practice and policy. International Journal of Information Management 57, 101994.

Ehrler, V., Hebes, P., 2012. Electromobility for city logistics—The solution to urban collapse? An analysis beyond theory. Social and Behavioral Sciences 48, 786–795.

Eiben, Á.E., Ellers, J., Meynen, G., Nyholm, S., 2021. Robot Evolution: Ethical Concerns. Frontiers in Robotics and AI 8, 344.

Eisenhardt, K.M., 1989. Building Theories from Case Study Research. Academy of Management Review 14 (4), 532–550.

ELA, 2023. European Logistics Association <https://www.elalog.eu/> (accessed 18 October 2023).

Elkington, J., 1994. Towards the sustainable corporation: Win-win-win business strategies for sustainable development. California management review 36 (2), 90–100.

Elkington, J., 1997. The triple bottom line. Environmental Management: Readings and Cases 2, pp. 49–66.

Elkington, J., Rowlands, I.H., 1999. Cannibals with forks: The triple bottom line of 21st century business. Alternatives Journal 25 (4).

Emerson, R.M., 1976. Social Exchange Theory. Annual Review of Sociology 2, 335–362.

Engels, B., Grunewald, M., 2017. Das Privacy Paradox: Digitalisierung versus Privatsphäre. Digitalisierung versus Privatsphäre 57, IW Kurzbericht.

EU Commission, 2020. White Paper on Artificial Intelligence—A European approach to excellence and trust. Brussels, 1–26.

European Commission, 2007. Green Paper: Towards a new culture of urban mobility. Comission of the European Communities <https://eur-lex.europa.eu/legal-content/EN/TXT/PDF/?uri=CELEX:52007DC0551&from=EN;%20last>.

Evans, J.S.B.T., 2008. Dual-processing accounts of reasoning, judgment, and social cognition. Annual Review of Psychology 59, 255–278.

Fabre, R.G., Camacho Ibáñez, J., Tejedor Escobar, P., 2021. Moral control and ownership in AI systems. AI & Soc 36 (1), 289–303.

Fan, J., Zhang, P., Yen, D.C., 2014. G2G information sharing among government agencies. Information & Management 51 (1), 120–128.

Farisco, M., Evers, K., Salles, A., 2020. Towards Establishing Criteria for the Ethical Analysis of Artificial Intelligence. Sci Eng Ethics 26 (5), 2413–2425.

Featherman, M., Fuller, M., 2003. Applying TAM to e-services adoption: the moderating role of perceived risk. 36th Annual Hawaii International Conference, 11.

Felzmann, H., Villaronga, E.F., Lutz, C., Tamò-Larrieux, A., 2019. Transparency you can trust: Transparency requirements for artificial intelligence between legal norms and contextual concerns. Big Data & Society 6 (1), 205395171986054.

Fenech, M.E., Buston, O., 2020. AI in Cardiac Imaging: A UK-Based Perspective on Addressing the Ethical, Social, and Political Challenges. Frontiers in cardiovascular medicine 7, 54.

Feng, L., Zhang, H., Lou, L., Chen, Y., 2018. A Blockchain-Based Collocation Storage Architecture for Data Security Process Platform of WSN. 2018 IEEE 22nd International Conference on Computer Supported Cooperative Work in Design ((CSCWD)), 75–80.

Feng, X.-Q., Pan, B.-H., 2021. The evolution of patent system: invention created by artificial intelligence. Procedia Computer Science 183, 245–253.

Fernandes, P.M., Santos, F.C., Lopes, M., 2020. Norms for beneficial A.I.: A computational analysis of the societal value alignment problem. AIC 33 (3–6), 155–171.

Fieser, J., 2000. Metaethics, Normative Ethics, and Applied Ethics: Historical and Contemporary Readings. Wadsworth/Thomson Learning.

Fiske, A., Henningsen, P., Buyx, A., 2019. Your Robot Therapist Will See You Now: Ethical Implications of Embodied Artificial Intelligence in Psychiatry, Psychology, and Psychotherapy. Journal of Medical Internet Research 21 (5), 1–12.

Flavián, C., Guinaliu, M., 2006. Consumer trust, perceived security and privacy policy: three basic elements of loyalty to a web site. Industrial Management & Data Systems 106 (5), 601–620.

Fleischmann, M., Ivens, B., 2019. Exploring the Role of Trust in Blockchain Adoption: An Inductive Approach. Proceedings of the 52nd Hawaii International Conference on System Sciences.

Floridi, L., 2018. Soft ethics, the governance of the digital and the General Data Protection Regulation. Philosophical transactions. Series A, Mathematical, physical, and engineering sciences 376 (2133).

Floridi, L., Cowls, J., Beltrametti, M., Chatila, R., Chazerand, P., Dignum, V., Luetge, C., Madelin, R., Pagallo, U., Rossi, F., Schafer, B., Valcke, P., Vayena, E., 2018. AI4People-An Ethical Framework for a Good AI Society: Opportunities, Risks, Principles, and Recommendations. Minds & Machines 28 (4), 689–707.

Floridi, L., Cowls, J., King, T.C., Taddeo, M., 2021. How to Design AI for Social Good: Seven Essential Factors, in: , Ethics, Governance, and Policies in Artificial Intelligence. Springer, Cham, 125–151.

Foa, U.G., Foa, E.B., 1974. Societal structures of the mind. Charles C Thomas.

Ford Media Center, 2023. Ford elektrisiert DHL-Flotte mit neuen E-Transit-Modellen <https://media.ford.com/content/fordmedia/feu/ch/de/news/2023/03/07/ford-elektrisiert-dhl-flotte-mit-neuen-e-transit-modellen.html#:~:text=Damit%20erzielt%20die%20elektrische%20Version,auch%20mit%20Wechselstrom%20geladen%20werden.> (accessed 19 October 2023).

Formosa, P., Ryan, M., 2021. Making moral machines: why we need artificial moral agents. AI & Soc 36 (3), 839–851.

Fornell, C., Larcker, D.F., 1981. Evaluating Structural Equation Models with Unobservable Variables and Measurement Error. Journal of Marketing Research 18 (1), 39–50.

Frank, D.-A., Chrysochou, P., Mitkidis, P., 2023. The paradox of technology: Negativity bias in consumer adoption of innovative technologies. Psychology & Marketing 40 (3), 554–566.

Frank, D.-A., Chrysochou, P., Mitkidis, P., Ariely, D., 2019. Human decision-making biases in the moral dilemmas of autonomous vehicles. Sci Rep 9 (1), 13080.

Franzke, A.S., Muis, I., Schäfer, M.T., 2021. Data Ethics Decision Aid (DEDA): a dialogical framework for ethical inquiry of AI and data projects in the Netherlands. Ethics Inf Technol, 1–17.

Fritz, O., Weber, C., König, A., Wolf, J., 2019. Ethische Aspekte der Künstlichen Intelligenz. KCT Schriftenreihe der FOM (1).

Fröding, B., Peterson, M., 2020. Friendly AI. Ethics Inf Technol, 1–8.

Fukuda-Parr, S., Gibbons, E., 2021. Emerging Consensus on 'Ethical AI': Human Rights Critique of Stakeholder Guidelines. Glob Policy 12 (S6), 32–44.

Gabriel, I., 2020. Artificial Intelligence, Values, and Alignment. Minds & Machines 30 (3), 411–437.

Gallin, S., Portes, A., 2024. Online shopping: How can algorithm performance expectancy enhance impulse buying? Journal of Retailing and Consumer Services 81, 103988.

Gamez, P., Shank, D.B., Arnold, C., North, M., 2020. Artificial virtue: the machine question and perceptions of moral character in artificial moral agents. AI & Soc 35 (4), 795–809.

Gao, L., Bai, X., 2014. A unified perspective on the factors influencing consumer acceptance of internet of things technology. Asia Pacific Journal of Marketing and Logistics 26 (2), 211–231.

Gefen, Karahanna, Straub, 2003. Trust and TAM in Online Shopping: An Integrated Model. MIS Quarterly 27 (1), 51.

Geis, J.R., Brady, A.P., Wu, C.C., Spencer, J., Ranschaert, E., Jaremko, J.L., Langer, S.G., Kitts, A.B., Birch, J., Shields, W.F., van Hoven Genderen, R., Kotter, E., Gichoya, J.W., Cook, T.S., Morgan, M.B., an Tang, Safdar, N.M., Kohli, M., 2019. Ethics of Artificial Intelligence in Radiology: Summary of the Joint European and North American Multisociety Statement. Canadian Association of Radiologists journal = Journal l'Association canadienne des radiologistes 70 (4), 329–334.

Gentile, C., Spiller, N., Noci, G., 2007. How to Sustain the Customer Experience: An Overview of Experience Components that Co-Create Value With the Customer. European Management Journal 25 (5), 395–410.

Gerdes, A., 2022. A participatory data-centric approach to AI Ethics by Design. Applied Artificial Intelligence 36 (1).

Ghasemaghaei, M., Hassanein, K., Benbasat, I., 2019. Assessing the design choices for online recommendation agents for older adults. MIS Quarterly 43 (1), 329–346.

Ghotbi, N., Ho, M.T., Mantello, P., 2021. Attitude of college students towards ethical issues of artificial intelligence in an international university in Japan. AI & Soc, 1–8.

Gibbons, E.D., 2021. Toward a More Equal World: The Human Rights Approach to Extending the Benefits of Artificial Intelligence. IEEE Technol. Soc. Mag. 40 (1), 25–30.

Gibert, M., Martin, D., 2021. In search of the moral status of AI: why sentience is a strong argument. AI & Soc, 1–12.
Gil-Cordero, E., Cabrera-Sánchez, J.P., Arrás-Cortés, M.J., 2020. Cryptocurrencies as a Financial Tool: Acceptance Factors. Mathematics 8 (11), 1974.
Gill, T., 2020. Blame It on the Self-Driving Car: How Autonomous Vehicles Can Alter Consumer Morality. SSRN Journal.
Glanemann, N., Willner, S.N., Levermann, A., 2020. Paris Climate Agreement passes the cost-benefit test. Nature Communications 11, 110.
Glaser, F., Zimmermann, K., Haferkorn, M., Weber, M.C., Siering, M., 2014. Bitcoin—Asset or Currency? Revealing Users' Hidden Intentions. Proceedings of the 22nd European Conference on Information Systems, Tel Aviv.
Globisch, J., Dütschke, E., Schneider Uta, 2013. Acceptance of electric vehicles by commercial users in the electric mobility pilot regions in Germany. ECEE Summer Transport and Mobility: How to Deliver Energy Efficiency, 973–983.
Godoe, P., Johansen, T.S., 2012. Understanding adoption of new technologies: Technology readiness and technology acceptance as an integrated concept. Journal of European Psychology Students 3, 38–52.
Gogoshin, D.L., 2021. Robot Responsibility and Moral Community. Frontiers in Robotics and AI 8.
Goh, C.S., Chong, H.Y., Jack, L., Faris, A.F.M., 2020. Revisiting triple bottom line within the context of sustainable construction: A systematic review. Journal of Cleaner Production 252, 119884.
Goirand, M., Austin, E., Clay-Williams, R., 2021. Implementing Ethics in Healthcare AI-Based Applications: A Scoping Review. Science and Engineering Ethics 27 (5), 61.
Gonçalves, L., Patrício, L., Grenha Teixeira, J., Wünderlich, N.V., 2020. Understanding the customer experience with smart services. Journal of Service Management 31 (4), 723–744.
Gong, X., Zhang, K.Z., Chen, C., Cheung, C.M., Lee, M.K., 2020. What drives trust transfer from web to mobile payment services? The dual effects of perceived entitativity. Information & Management 57 (7), 103250.
Gordon, J.-S., 2020. Building Moral Robots: Ethical Pitfalls and Challenges. Sci Eng Ethics 26 (1), 141–157.
Gordon, J.-S., 2021. Artificial moral and legal personhood. AI & Soc 36 (2), 457–471.
Gordon, J.-S., Gunkel, D.J., 2022. Moral Status and Intelligent Robots. The Southern Journal of Philosophy 60 (1), 88–117.
Gottselig, V., Wuppermann, A., Herrmann, C., 2023. Effects of green nudges on consumer valuation of sustainable food: A discrete choice experiment. Ecological Perspectives for Science and Society 32 (2), 233–240.
Gourville, J.T., 2006. Eager sellers and stony buyers: Understanding the psychology of new-product adoption. Harvard Business Review 84 (6), 98–106.
Grand View Research, 2019. Smart Refrigerators Market Size Analysis Report By Door Type (Single, Double, Side by Side, French), By End user (Residential, Commercial) By Distribution Channel (Offline, Online), And Segment Forecasts, 2019–2025, San Francisco.
Grewal, D., Roggeveen, A.L., 2020. Understanding Retail Experiences and Customer Journey Management. Journal of Retailing 96 (1), 3–8.

Griskevicius, V., Tybur, J., van den Bergh, B., 2010. Going green to be seen: status, reputation, and conspicuous conservation. Journal of Personality and Social Psychology 98 (3), 392.

Groot, J.I.M. de, Steg, L., 2009. Morality and prosocial behavior: the role of awareness, responsibility, and norms in the norm activation model. The Journal of social psychology 149 (4), 425–449.

Gruson, D., Helleputte, T., Rousseau, P., Gruson, D., 2019. Data science, artificial intelligence, and machine learning: Opportunities for laboratory medicine and the value of positive regulation. Clinical Biochemistry 69, 1–7.

Guha, A., Grewal, D., Kopalle, P.K., Haenlein, M., Schneider, M.J., Jung, H., Moustafa, R., Hegde, D.R., Hawkins, G., 2021. How artificial intelligence will affect the future of retailing. Journal of Retailing 97 (1), 28–41.

Guha, I., 2023. Total Cost Of Online Returns Breaks $200 Billion—Can Retail Afford It? <https://www.forbes.com/councils/forbescommunicationscouncil/2023/10/03/total-cost-of-online-returns-breaks-200-billion-can-retail-afford-it/> (accessed 25 October 2024).

Gupta, M., Parra, C.M., Dennehy, D., 2021a. Questioning Racial and Gender Bias in AI-based Recommendations: Do Espoused National Cultural Values Matter? Inf Syst Front, 1–17.

Gupta, S., Giri, S., Srivastava, T., Agarwal, P., Sharma, R., Agrawal, A., 2021b. Smart Refrigerator based on 'Internet of Things'. 2021 International Conference on Advance Computing and Innovative Technologies in Engineering (ICACITE), 436–439.

Ha, S., Stoel, L., 2012. Online apparel retailing: roles of e-shopping quality and experiential e-shopping motives. Journal of Service Management 23 (2).

Haas, J., 2020. Moral Gridworlds: A Theoretical Proposal for Modeling Artificial Moral Cognition. Minds & Machines 30 (2), 219–246.

Haenlein, M., Kaplan, A., 2019. A Brief History of Artificial Intelligence: On the Past, Present, and Future of Artificial Intelligence. California management review 61 (4), 5–14.

Hagendorff, T., 2020. The ethics of AI ethics: An evaluation of guidelines. Minds and Machines 30 (1), 99–120.

Hair, J.F., Black, W.C., Babin, B.J., Anderson, R.E., 2014. Multivariate data analysis, Harlow, Essex.

Hair, J.F., Black, W.C., Babin, B.J., Anderson, R.E., 2019. Multivariate data analysis, Eighth edition ed. Cengage, Andover, Hampshire.

Hair, J.F., JR, Sarstedt, M., Hopkins, L., Kuppelwieser, V.G., 2014. Partial Least Squares Structural Equation Modeling (PLS-SEM): An Emerging Tool in Business Research. European Business Review 26 (2), 106–121.

Halpern, D., 2015. Inside the nudge unit: How small changes can make a big difference. W.H. Allen, London.

Hamilton, R.H., Davison, H.K., 2021. Legal and Ethical Challenges for HR in Machine Learning. Employ Respons Rights J, 1–21.

Handelsblatt, 2020. Paypal drängt in die Welt der Krypto-Währungen—Bitcoin-Kurs steigt auf Jahreshoch <https://www.handelsblatt.com/finanzen/banken-versicherungen/zahlungsanbieter-paypal-draengt-in-die-welt-der-krypto-waehrungen-bitcoinkurs-steigt-auf-jahresrekord/26295380.html> (accessed 21 August 2023).

Handgraaf, M.J., van Lidth de Jeude, M.A., Appelt, K.C., 2013. Public praise vs. private pay: Effects of rewards on energy conservation in the workplace. Ecological Economics 86, 86–92.

Handrich, M., 2021. Alexa, You Freak Me Out–Identifying Drivers of Innovation Resistance and Adoption of Intelligent Personal Assistants. Forty-Second International Conference on Information Systems, Austin.

Hankammer, S., Kleer, R., Piller, F.T., 2021. Sustainability nudges in the context of customer co-design for consumer electronics. Journal of Business Economics 91 (6), 897–933.

Hansen, P.G., Jespersen, A.M., 2013. Nudge and the manipulation of choice: A framework for the responsible use of the nudge approach to behaviour change in public policy. European Journal of Risk Regulation 4 (1), 3–28.

Hardman, S., Jenn, A., Tal, G., Axsen, J., Beard, G., Daina, N., Figenbaum, E., Jakobsson, N., Jochem, P., Kinnear, N., Plötz, P., Pontes, J., Refa, N., Sprei, F., Turrentine, T., Witkamp, B., 2018. A review of consumer preferences of and interactions with electric vehicle charging infrastructure. Transportation Research Part D: Transport and Environment 62, 508–523.

Hassan, S.M., Rahman, Z., Paul, J., 2022. Consumer ethics: A review and research agenda. Psychology & Marketing 39, 111–130.

Häubl, G., Trifts, V., 2000. Consumer Decision Making in Online Shopping Environments: The Effects of Interactive Decision Aids. Marketing Science 19 (1), 4–21.

Hauslbauer, A.L., Schade, J., Drexler, C.E., Petzoldt, T., 2022. Extending the theory of planned behavior to predict and nudge toward the subscription to a public transport ticket. European Transport Research Review 14 (1), 1–14.

Heatherton, T.F., Polivy, J., 1991. Development and validation of a scale for measuring state self-esteem. Journal of Personality and Social Psychology 60 (6), 895–910.

Héder, M., 2020. The epistemic opacity of autonomous systems and the ethical consequences. AI & Soc, 1–9.

Heidenreich, S., Handrich, M., 2015a. Adoption of Technology-Based Services: The Role of Customers' Willingness to Co-Create. Journal of Service Management 26 (1), 44–71.

Heidenreich, S., Handrich, M., 2015b. What about passive innovation resistance? Investigating adoption-related behavior from a resistance perspective. Journal of Product Innovation Management 32 (6), 878–903.

Heidenreich, S., Spieth, P., 2013. Why innovations fail—The case of passive and active innovation resistance. International Journal of Innovation Management 17 (5), 1350021.

Henn, N., Lohwasser, T.S., 2020. The advances of community cloud computing in the business-to-business buying process. Diskussionspapier des Instituts für Organisationsökonomik 6/2020. Münster: Westfälische Wilhelms-Universität Münster, Institut für Organisationsökonomik, Münster <https://www.econstor.eu/handle/10419/222484>.

Hennink, M., Kaiser, B.N., 2022. Sample sizes for saturation in qualitative research: A systematic review of empirical tests. Social science & Medicine 292, 114523.

Hermann, E., 2021a. Artificial intelligence and mass personalization of communication content—An ethical and literacy perspective. New Media & Society, 146144482110227.

Hermann, E., 2021b. Leveraging Artificial Intelligence in Marketing for Social Good-An Ethical Perspective. J Bus Ethics, 1–19.

Herzenstein, M., Posavac, S.S., Brakus, J.J., 2007. Adoption of New and Really New Products: The Effects of Self-Regulation Systems and Risk Salience. Journal of Marketing Research 44 (2), 251–260.

Herzog, C., 2021. Three Risks That Caution Against a Premature Implementation of Artificial Moral Agents for Practical and Economical Use. Sci Eng Ethics 27 (1), 3.

Heyder, T., Passlack, N., Posegga, O., 2023. Ethical management of human-AI interaction: Theory development review. The Journal of Strategic Information Systems 32 (3), 101772.

Hickman, E., Petrin, M., 2021. Trustworthy AI and Corporate Governance: The EU's Ethics Guidelines for Trustworthy Artificial Intelligence from a Company Law Perspective. European Business Organization Law Review 22 (4), 593–625.

Hill, J., Lee, H.-H., 2012. Young Generation Y consumers' perceptions of sustainability in the apparel industry. Journal of Fashion Marketing and Management 16 (4), 477–491.

Hilton, D., Charalambides, L., Demarque, C., Waroquier, L., Raux, C., 2014. A tax can nudge: The impact of an environmentally motivated bonus/malus fiscal system on transport preferences. Journal of Economic Psychology 42, 17–27.

Hirschman, E.C., 1980. Innovativeness, novelty seeking, and consumer creativity. Journal of Consumer Research 7 (3), 283–295.

Hirschman, E.C., Holbrook, M.B., 1982. Hedonic Consumption: Emerging Concepts, Methods and Propositions. Journal of Marketing 46 (3), 92–101.

Ho, C.W.L., Ali, J., Caals, K., 2020. Ensuring trustworthy use of artificial intelligence and big data analytics in health insurance. Bulletin of the World Health Organization 98 (4), 263–269.

Hoffman, D.L., Moreau, C.P., Stremersch, S., Wedel, M., 2022. The Rise of New Technologies in Marketing: A Framework and Outlook. Journal of Marketing 86 (1), 1–6.

Hohenstein, J., Jung, M., 2020. AI as a moral crumple zone: The effects of AI-mediated communication on attribution and trust. Computers in Human Behavior 106, 106190.

Hollebeek, L.D., Sprott, D.E., Brady, M.K., 2021. Rise of the Machines? Customer Engagement in Automated Service Interactions. Journal of Service Research 24 (1), 3–8.

Holub, M., Johnson, J., 2018. Bitcoin research across disciplines. The Information Society 34 (2), 114–126.

Homans, G.C., 1958. Social Behavior as Exchange. American Journal of Sociology 63 (6), 597–606.

Hong, D., Cho, C.-H., 2023. Factors Affecting Innovation Resistance of Smartphone AI Voice Assistants. International Journal of Human–Computer Interaction 39 (13), 2557–2572.

Hong, I.B., Cha, H.S., 2013. The mediating role of consumer trust in an online merchant in predicting purchase intention. International Journal of Information Management 33 (6), 927–939.

Hong, J.-W., Choi, S., Williams, D., 2020. Sexist AI: An Experiment Integrating CASA and ELM. International Journal of Human–Computer Interaction 36 (20), 1928–1941.

Hong, Y., Pavlou, P.A., 2014. Product Fit Uncertainty in Online Markets: Nature, Effects, and Antecedents. Information Systems Research 25 (2), 328–344.

Hooftman, N., Messagie, M., van Mierlo, J., Coosemans, T., 2018. A review of the European passenger car regulations—Real driving emissions vs local air quality. Renewable and Sustainable Energy Reviews 86, 1–21.

Hossain, S., Abdelgawad, A., 2018. Smart Refrigerator Based on Internet of Things (IoT): An Approach of Efficient Food Management. Proceedings of the 2nd International Conference on Smart Digital Environment.

Hoyer, W.D., Kroschke, M., Schmitt, B., Kraume, K., Shankar, V., 2020. Transforming the Customer Experience Through New Technologies. Journal of Interactive Marketing 51, 57–71.

Huang, B.I., 2019. Law's Halo and the Moral Machine. Columbia Law Review 119, 1811–1828.

Huang, D., Jin, X., Coghlan, A., 2021. Advances in consumer innovation resistance research: A review and research agenda. Technological Forecasting and Social Change 166, 120594.

Huang, M.-H., Rust, R.T., 2021. A strategic framework for artificial intelligence in marketing. Journal of the Academy of Marketing Science 49 (1), 30–50.

Huang, N., Qin, G., 2011. A study of online virtual fitting room adoption based on UTAUT. International Conference on E-Business and E-Government (ICEE). IEEE, 1–4.

Hughes, L., Dwivedi, Y.K., Misra, S.K., Rana, N.P., Raghavan, V., Akella, V., 2019. Blockchain research, practice and policy: Applications, benefits, limitations, emerging research themes and research agenda. International Journal of Information Management 49, 114–129.

Hummel, D., Maedche, A., 2019. How effective is nudging? A quantitative review on the effect sizes and limits of empirical nudging studies. Journal of Behavioral and Experimental Economics 80, 47–58.

Hynek, N., Solovyeva, A., 2021. Operations of power in autonomous weapon systems: ethical conditions and socio-political prospects. AI & Soc 36 (1), 79–99.

Ibáñez, J.C., Olmeda, M.V., 2021. Operationalising AI ethics: how are companies bridging the gap between practice and principles? An exploratory study. AI & SOCIETY, 1–25.

IBM, 2024. What is AI ethics? <https://www.ibm.com/topics/ai-ethics> (accessed 15 January 2024).

IEA, 2021. Net Zero by 2050: A Roadmap for the Global Energy Sector, 4th ed. International Energy Agency.

IISD, 1995. Oslo roundtable on sustainable production and consumption. The International Institute for Sustainable Development, New York.

Iphofen, R., Kritikos, M., 2021. Regulating artificial intelligence and robotics: ethics by design in a digital society. Contemporary Social Science 16 (2), 170–184.

Ivanov, A., Mou, Y., Tawira, L., 2022. Avatar personalisation vs. privacy in a virtual try-on app for apparel shopping. International Journal of Fashion Design, Technology and Education, 1–10.

Iwan, S., Nürnberg, M., Jedlinski, M., Kijewska, K., 2021. Efficiency of light electric vehicles in last mile deliveries—Szczecin case study. Sustainable Cities and Society.

Jackisch, K.-R., 2021. Was der digitale Euro können soll <https://www.tagesschau.de/wirtschaft/ezb-digitaler-euro-101.html> (accessed 21 August 2023).

Jamali, D., 2006. Insights into triple bottom line integration from a learning organization perspective. Bsuiness Process Management 12 (6), 809–821.

Janssen, M., Brous, P., Estevez, E., Barbosa, L.S., Janowski, T., 2020. Data governance: Organizing data for trustworthy Artificial Intelligence. Government Information Quarterly 37 (3), 101493.

Jansson, J., Nordlund, A., Westin, K., 2017. Examining drivers of sustainable consumption: The influence of norms and opinion leadership on electric vehicle adoption in Sweden. Journal of Cleaner Production 154, 176–187.

Javaria, K., Masood, O., Garcia, F., 2020. Strategies to manage the risks faced by consumers in developing e-commerce. Insights into Regional Development 2 (4), 774–783.

Jiang, F., Ma, L., Broyd, T., Li, J., Jia, J., Luo, H., 2023. Systematic framework for sustainable urban road alignment planning. Transportation Research Part D: Transport and Environment 120, 103796.

Jobin, A., Ienca, M., Vayena, E., 2019. The global landscape of AI ethics guidelines. Nature Machine Intelligence 1 (9), 389–399.

Jobson, D., Mar, V., Freckelton, I., 2022. Legal and ethical considerations of artificial intelligence in skin cancer diagnosis. The Australasian journal of dermatology 63 (1).

Joerin, A., Rauws, M., Fulmer, R., Black, V., 2020. Ethical Artificial Intelligence for Digital Health Organizations. Cureus 12 (3), e7202.

John, P., Blume, T., 2018. How best to nudge taxpayers? The impact of message simplification and descriptive social norms on payment rates in a central London local authority. Journal of Behavioral Public Administration 1 (1), 1–11.

Jörling, M., Böhm, R., Paluch, S., 2019. Service Robots: Drivers of Perceived Responsibility for Service Outcomes. Journal of Service Research 22 (4), 404–420.

Jotterand, F., Bosco, C., 2020. Keeping the "Human in the Loop" in the Age of Artificial Intelligence : Accompanying Commentary for "Correcting the Brain?" by Rainey and Erden. Sci Eng Ethics 26 (5), 2455–2460.

Juan, A., Mendez, C., Faulin, J., Armas, J. de, Grasman, S., 2016. Electric Vehicles in Logistics and Transportation: A Survey on Emerging Environmental, Strategic, and Operational Challenges. Energies 9 (2), 86.

Kahlert, M., Constantinides, E., Vries, S.A. de, 2017. The relevance of technological autonomy in the customer acceptance of IoT services in retail. Proceedings of the Second International Conference on Internet of things, Data and Cloud Computing (12), 1–7.

Kalinic, Z., Marinkovic, V., Molinillo, S., Liébana-Cabanillas, F., 2019. A multi-analytical approach to peer-to-peer mobile payment acceptance prediction. Journal of Retailing and Consumer Services 49, 143–153.

Kalpoe, R., 2020. Technology acceptance and return management in apparel e-commerce. Journal of Supply Chain Management Science 1 (3–4), 118–137.

Kane, T.B., 2019. Artificial Intelligence in Politics: Establishing Ethics. IEEE Technol. Soc. Mag. 38 (1), 72–80.

Kang, H., Jung, E.H., 2021. The smart wearables-privacy paradox: A cluster analysis of smartwatch users. Behaviour & Information Technology 40 (16), 1755–1768.

Kant, I., 1785. Groundwork of the metaphysics of morals. J. F. Hartknoch.

Kaplan, A., Haenlein, M., 2019. Siri, Siri, in my hand: Who's the fairest in the land? On the interpretations, illustrations, and implications of artificial intelligence. Business Horizons 62 (1), 15–25.

Kaplan, B., Maxwell, J.A., 2005. Qualitative research methods for evaluating computer information systems. Evaluating the organizational impact of healthcare information systems, Springer, New York, 30–55.

Kaplan, S., Gruber, J., Reinthaler, M., Klauenberg, J., 2016. Intentions to introduce electric vehicles in the commercial sector: A model based on the theory of planned behaviour. Research in Transportation Economics 55 (6), 12–19.

Karame, G.O., Androulaki, E., Capkun, S., 2012. Double-spending fast payments in bitcoin. Proceedings of the ACM conference on Computer and communications security.

Kaur, P., Dhir, A., Singh, N., Sahu, G., Almotairi, M., 2020. An innovation resistance theory perspective on mobile payment solutions. Journal of Retailing and Consumer Services 55, 102059.

KE Consult, 2020. KEP Studie 2020 Analyse des Markets in Deutschland. Bundesverband Paket & Expresss Logistik.

KE-Consult, 2023. Perspektiven eröffnen, Gemeinschaft gestalten: KEP-Studie 2023—Analyse des Markets in Deutschland. Bundesverband Paket und Expresslogistik e. V., Berlin.

Keen, P., Ballance, G., Chan, S., Schrump, S., 1999. Electronic commerce relationships: Trust by design, Prentice Hall PTR.

Kelley, D., Atreides, K., 2020. AGI Protocol for the Ethical Treatment of Artificial General Intelligence Systems. Procedia Computer Science 169, 501–506.

Kelley, D., Twyman, M., 2020. Biasing in an Independent Core Observer Model Artificial General Intelligence Cognitive Architecture. Procedia Computer Science 169, 535–541.

Kelly, S., Kaye, S.-A., Oviedo-Trespalacios, O., 2023. What factors contribute to the acceptance of artificial intelligence? A systematic review. Telematics and Informatics 77 77, 101925.

Kempt, H., Nagel, S.K., 2022. Responsibility, second opinions and peer-disagreement: ethical and epistemological challenges of using AI in clinical diagnostic contexts. Journal of medical ethics 48 (4), 222–229.

Kenny, L.M., Nevin, M., Fitzpatrick, K., 2021. Ethics and standards in the use of artificial intelligence in medicine on behalf of the Royal Australian and New Zealand College of Radiologists. Journal of medical imaging and radiation oncology 65 (5), 486–494.

Kerasidou, A., 2021. Ethics of artificial intelligence in global health: Explainability, algorithmic bias and trust. Journal of oral biology and craniofacial research 11 (4), 612–614.

Kerr, A., Barry, M., Kelleher, J.D., 2020. Expectations of artificial intelligence and the performativity of ethics: Implications for communication governance. Big Data & Society 7 (1), 205395172091593.

Khan, I.S., Ahmad, M.O., Majava, J., 2021. Industry 4.0 and sustainable development: A systematic mapping of triple bottom line, Circular Economy and Sustainable Business Models perspectives. Journal of Cleaner Production 297, 126655.

Khargonekar, P.P., Sampath, M., 2020. A Framework for Ethics in Cyber-Physical-Human Systems. IFAC-PapersOnLine 53 (2), 17008–17015.

Khrais, L.T., 2020. Role of Artificial Intelligence in Shaping Consumer Demand in E-Commerce. Future Internet 12 (12), 226.

Kieras, D., Polson, P.G., 1985. An approach to the formal analysis of user complexity. International Journal of Man-Machine Studies 22 (4), 365–394.

Kim, C., Costello, F.J., Lee, K.C., 2020. The Unobserved Heterogeneneous Influence of Gamification and Novelty-Seeking Traits on Consumers' Repurchase Intention in the Omnichannel Retailing. Frontiers in Psychology 11, 1664.

Kim, D., Park, K., Park, Y., Ahn, J.-H., 2019. Willingness to provide personal information: Perspective of privacy calculus in IoT services. Computers in Human Behavior 92, 273–281.

Kim, K., Moon, S., Lee, J., Kim, J., 2023. Environmental impacts of transportation policy and vehicle market share predictions. Transportation Research Part D: Transport and Environment, 103550.

Kim, M.-J., Chung, N., Lee, C.-K., 2011. The effect of perceived trust on electronic commerce: Shopping online for tourism products and services in South Korea. Tourism Management 32 (2), 256–265.

King, O.C., 2020. Presumptuous aim attribution, conformity, and the ethics of artificial social cognition. Ethics Inf Technol 22 (1), 25–37.

Kleijnen, M., Lee, N., Wetzels, M., 2009. An exploration of consumer resistance to innovation and its antecedents. Journal of Economic Psychology 30, 344–357.

Klein, J.G., Smith, N.C., John, A., 2004. Why We Boycott: Consumer Motivations for Boycott Participation. Journal of Marketing 68 (3), 92–109.

Köbis, N., Bonnefon, J.-F., Rahwan, I., 2021. Bad machines corrupt good morals. Nat Hum Behav 5 (6), 679–685.

Kochupillai, M., Lütge, C., Poszler, F., 2020. Programming Away Human Rights and Responsibilities? "The Moral Machine Experiment" and the Need for a More "Humane" AV Future. Nanoethics 14 (3), 285–299.

Koh, T.K., Fichman, M., Kraut, R.E., 2012. Trust Across Borders: Buyer-Supplier Trust in Global Business-to-Business E-Commerce. Journal of the association for information systems 13 (11), 886–922.

Komiak, S.Y.X., Benbasat, I., 2006. The Effects of Personalization and Familarity on Trust and Adoption of Recommendation Agents. MIS Quarterly 30 (4), 941.

Koufaris, M., Hampton-Sosa, W., 2004. The development of initial trust in an online company by new customers. Information & Management 41 (3), 377–397.

KPMG LLP, 2023. The technology we can't live without. KPMG Consumer Technology Survey.

Kretzschmar, K., Tyroll, H., Pavarini, G., Manzini, A., Singh, I., 2019. Can Your Phone Be Your Therapist? Young People's Ethical Perspectives on the Use of Fully Automated Conversational Agents (Chatbots) in Mental Health Support. Biomedical Informatics Insights 11, 1178222619829083.

Kriebitz, A., Lütge, C., 2020. Artificial Intelligence and Human Rights: A Business Ethical Assessment. Bus. and hum. rights j. 5 (1), 84–104.

Krijger, J., 2021. Enter the metrics: critical theory and organizational operationalization of AI ethics. AI & SOCIETY.

Kristoufek, L., 2015. What are the main drivers of the Bitcoin price? Evidence from wavelet coherence analysis. PLOS ONE 10 (4), e0123923.

Kroll, T., Paukstadt, U., Kreidermann, K., Mirbabaeie, M., 2019. Nudging People to Save Energy in Smart Homes with Social Norms and Self-Commitment. Proceedings of the 2019 European Conference on Information Systems, Stockholm, Sweden.

Kuchenbecker, M., Manner-Romberg, H., Zimmermann, J., 2021. Logistik auf der letzten Meile: Reallabor Stadt, 1st ed. logix.

Kuebel, H., Zarnekow, R., 2015. Exploring platform adoption in the smart home case. Proceedings of the International Conference on Information Systems (ICIS) 15.

Kuehnl, C., Jozic, D., Homburg, C., 2019. Effective customer journey design: consumers' conception, measurement, and consequences. Journal of the Academy of Marketing Science 47 (3), 551–568.

Kumar, V., Ramachandran, D., Kumar, B., 2021. Influence of new-age technologies on marketing: A research agenda. Journal of Business Research 125, 864–877.

Kumar Jain, N., Bhaskar, K., Jain, S., 2022. What drives adoption intention of electric vehicles in India? An integrated UTAUT model with environmental concerns, perceived risk and government support. Research in Transportation Business & Management 42, 100730.

Kushwaha, T., Shankar, V., 2013. Are Multichannel Customers Really more Valuable? The Moderating Role of Product Category Characteristics. Journal of Marketing 77 (4), 67–85.

Laakasuo, M., Palomäki, J., Köbis, N., 2021. Moral Uncanny Valley: A Robot's Appearance Moderates How its Decisions are Judged. Int J of Soc Robotics 13 (7), 1679–1688.

Lal, A., Renaldy, T., Breuning, L., Hamacher, T., You, F., 2023. Electrifying light commercial vehicles for last-mile deliveries: Environmental and economic perspectives. Journal of Cleaner Production, 137933.

Landon-Murray, M., Mujkic, E., Nussbaum, B., 2019. Disinformation in Contemporary U.S. Foreign Policy: Impacts and Ethics in an Era of Fake News, Social Media, and Artificial Intelligence. Public Integrity 21 (5), 512–522.

Lannon, C., Nelson, J., Cunneen, M., 2021. Ethical AI for Automated Bus Lane Enforcement. Sustainability 13 (21).

Lara, F., Deckers, J., 2020. Artificial Intelligence as a Socratic Assistant for Moral Enhancement. Neuroethics 13 (3), 275–287.

Laroche, M., Cleveland, M., Bergeron, J., Goutaland, C., 2003. The Knowledge-Experience-Evaluation Relationship: A Structural Equations Modeling Test of Gender Differences. Canadian Journal of Administrative Sciences / Revue Canadienne des Sciences de l'Administration 20 (3), 246–259.

Larson, D.B., Magnus, D.C., Lungren, M.P., Shah, N.H., Langlotz, C.P., 2020. Ethics of Using and Sharing Clinical Imaging Data for Artificial Intelligence: A Proposed Framework. Radiology 295 (3), 675–682.

Larsson, S., 2020. On the Governance of Artificial Intelligence through Ethics Guidelines. AsianJLS 7 (3), 437–451.

Laufer, R.S., Wolfe, M., 1977. Privacy as a Concept and a Social Issue: A Multidimensional Developmental Theory. Journal of Social Issues 33 (3), 22–42.

Laukkanen, T., 2016. Consumer adoption versus rejection decisions in seemingly similar service innovations: The case of the Internet and mobile banking. Journal of Business Research 69 (7), 2432–2439.

Laukkanen, T., Sinkkonen, S., Kivijärvi, M., Laukkanen, P., 2007. Innovation resistance among mature consumers. Journal of Consumer Marketing 24 (7), 419–427.

Lauwaert, L., 2021. Artificial intelligence and responsibility. AI & Soc 36 (3), 1001–1009.

Lebeau, P., Macharis, C., van Mierlo, J., 2016. Exploring the choice of battery electric vehicles in city logistics: A conjoint-based choice analysis. Transportation Research Part E: Logistics and Transportation Review 91 (4), 245–258.

Lee, A.Y., Aaker, J.L., 2004. Bringing the frame into focus: the influence of regulatory fit on processing fluency and persuasion. Journal of Personality and Social Psychology 86 (2), 205–218.

Lee, E.-J., Choi, H., Han, J., Kim, D.H., Ko, E., Kim, K.H., 2020. How to "Nudge" your consumers toward sustainable fashion consumption: An fMRI investigation. Journal of Business Research 117, 642–651.

Lee, H., Xu, Y., 2020. Classification of virtual fitting room technologies in the fashion industry: from the perspective of consumer experience. International Journal of Fashion Design, Technology and Education 13 (1), 1–10.

Lee, K.C., Kang, I., McKnight, D.H., 2007. Transfer From Offline Trust to Key Online Perceptions: An Empirical Study. IEEE Transactions on Engineering Management 54 (4), 729–741.

Leerkamp, B., Holthaus, T., Schlott, M., Klein, J., 2022. Comparative Study on Urban Logistics and Last Mile Delivery Processes in Germany and China. Deutsche Gesellschaft für Internationale Zusammenarbeit (GIZ) GmbH.

Lehner, M., Mont, O., Heiskanen, E., 2016. Nudging—A promising tool for sustainable consumption behaviour? Journal of Cleaner Production 134, 166–177.

Leotti, L.A., Iyengar, S.S., Ochsner, K.N., 2010. Born to choose: the origins and value of the need for control. Trends in Cognitive Sciences 14 (10), 457–463.

Lepri, B., Oliver, N., Pentland, A., 2021. Ethical machines: The human-centric use of artificial intelligence. iScience 24 (3), 102249.

Leung, E., Paolacci, G., Puntoni, S., 2018. Man versus machine: Resisting automation in identity-based consumer behavior. Journal of Marketing Research 55 (6), 818–831.

Li, H., Wu, J., Gao, Y., Shi, Y., 2016. Examining individuals' adoption of healthcare wearable devices: An empirical study from privacy calculus perspective. International Journal of Medical Informatics 88, 8–17.

Li, J., Abbasi, A., Cheema, A., Abraham, L.B., 2020. Path to purpose? How online customer journeys differ for hedonic versus utilitarian purchases. Journal of Marketing 84 (4), 127–146.

Li, O., 2021. Problems with "Friendly AI". Ethics Inf Technol, 1–8.

Li, X., Wang, C.A., 2017. The technology and economic determinants of cryptocurrency exchange rates: The case of Bitcoin. Decision Support Systems 95, 49–60.

Lian, J.-W., Yen, D.C., 2013. To buy or not to buy experience goods online: Perspective of innovation adoption barriers. Computers in Human Behavior 29 (3), 665–672.

Liao, F., Molin, E., van Wee, B., 2017. Consumer preferences for electric vehicles: a literature review. Transport Reviews 37 (3), 252–275.

Libai, B., Bart, Y., Gensler, S., Hofacker, C.F., Kaplan, A., Kötterheinrich, K., Kroll, E.B., 2020. Brave New World? On AI and the Management of Customer Relationships. Journal of Interactive Marketing 51, 44–56.

Lichtlé, M.-C., 2007. The effect of an advertisement's colour on emotions evoked by attitude towards the ad. International Journal of Advertising 26 (1), 37–62.

Lifewire, 2021. What Is a Smart Refrigerator? <https://www.lifewire.com/smart-refrigerator-4158327> (accessed 16 September 2022).

Lim, N., 2003. Consumers' perceived risk: sources versus consequences. Electronic Commerce Research and Applications 2 (3), 216–228.

Lindenberg, S., Steg, L., 2007. Normative, Gain and Hedonic Goal Frames Guiding Environmental Behavior. Journal of Social Issues 63 (1), 117–137.

Lindstad, E., Ask, T.O., Cariou, P., Eskeland, G.S., Rialland, A., 2023. Wise use of renewable energy in transport. Transportation Research Part D: Transport and Environment 119, 103713.

Liu, B., Miltgen, C.L., Xia, H., 2022. Disclosure decisions and the moderating effects of privacy feedback and choice. Decision Support Systems 155, 113717.

Liu, H., Yang, R., Wang, L., Liu, P., 2019a. Evaluating Initial Public Acceptance of Highly and Fully Autonomous Vehicles. International Journal of Human–Computer Interaction 35 (11), 919–931.

Liu, H.-W., Lin, C.-F., Chen, Y.-J., 2019b. Beyond State v Loomis: artificial intelligence, government algorithmization and accountability. International Journal of Law and Information Technology 27 (2), 122–141.

Liu, H.-Y., Maas, M.M., 2021. 'Solving for X?' Towards a problem-finding framework to ground long-term governance strategies for artificial intelligence. Futures 126, 102672.

Liu, R.X., Kuang, J., Gong, Q., Hou, X.L., 2003. Principal component regression analysis with SPSS. Computer Methods and Programs in Biomedicine 71 (2), 141–147.

Liu, Z., Ben, S., Zhang, R., 2019c. Factors affecting consumers' mobile payment behavior: a meta-analysis. Electronic Commerce Research 19 (3), 575–601.

Livingston, S., Risse, M., 2019. The Future Impact of Artificial Intelligence on Humans and Human Rights. Ethics int. aff. 33 (02), 141–158.

Loiacono, E., Watson, R.T., Goodhue, D.L., 2002. WebQual: A measure of website quality. Marketing Theory and Applications 13 (3), 432–438.

Longoni, C., Bonezzi, A., Morewedge, C.K., 2019. Resistance to Medical Artificial Intelligence. Journal of Consumer Research 46 (4), 629–650.

Longoni, C., Cian, L., 2022. Artificial Intelligence in Utilitarian vs. Hedonic Contexts: The "Word-of-Machine" Effect. Journal of Marketing 86 (1), 91–108.

Loureiro, S.M.C., Guerreiro, J., Tussyadiah, I., 2021. Artificial intelligence in business: State of the art and future research agenda. Journal of Business Research 129, 911–926.

Lu, H., Li, Y., Chen, M., Kim, H., Serikawa, S., 2018. Brain Intelligence: Go beyond Artificial Intelligence. Mobile Netw Appl 23 (2), 368–375.

Lu, Y., Yang, S., Chau, P.Y., Cao, Y., 2011. Dynamics between the trust transfer process and intention to use mobile payment services: A cross-environment perspective. Information & Management 48 (8), 393–403.

Lu, Y., Zhao, L., Wang, B., 2010. From virtual community members to C2C e-commerce buyers: Trust in virtual communities and its effect on consumers' purchase intention. Electronic Commerce Rcsearch and Applications 9 (4), 346–360.

Luk, J.W., Pruitt, L.D., Smolenski, D.J., Tucker, J., Workman, D.E., Belsher, B.E., 2021. From everyday life predictions to suicide prevention: Clinical and ethical considerations in suicide predictive analytic tools. Journal of Clinical Psychology.

Lukin, E., Krajnović, A., Bosna, J., 2022. Sustainability Strategies and Achieving SDGs: A Comparative Analysis of Leading Companies in the Automotive Industry. Sustainability 14 (7), 4000.

Lundblad, L., Davies, I.A., 2016. The values and motivations behind sustainable fashion consumption. Journal of Consumer Behaviour 15 (2), 149–162.

Mabaso, B.A., 2021. Computationally rational agents can be moral agents. Ethics Inf Technol 23 (2), 137–145.

MacIntyre, A., 1985. After Virtue. Duckworth, London.

Macnish, K., Fernandez Inguanzo, A., 2019. Case Study—Customer Relation Management, Smart Information Systems and Ethics. The ORBIT Journal 2 (2), 1–24.

Macnish, K., Fernandez Inguanzo, A., Kirichenko, A., 2019. Smart Information Systems in Cybersecurity. ORBIT Journal 2 (2), 1–26.

Magrani, E., 2019. New perspectives on ethics and the laws of artificial intelligence. Internet Policy Review 8 (3).

Mahmud, H., Islam, A.N., Mitra, R.K., 2023. What drivers managers towards algorithm aversion and how to overcome it? Mitigating the impact of innovation resistance through technology readiness. Technological Forecasting and Social Change 193, 122641.

Maity, M., Dass, M., 2014. Consumer decision-making across modern and traditional channels: E-commerce, m-commerce, in-store. Decision Support Systems 61, 34–46.

Maki, A., Burns, R.J., Ha, L., Rothman, A.J., 2016. Paying people to protect the environment: A meta-analysis of financial incentive interventions to promote proenvironmental behaviors. Journal of Environmental Psychology 47, 242–255.

Mantelero, A., 2018. AI and Big Data: A blueprint for a human rights, social and ethical impact assessment. Computer Law & Security Review 34 (4), 754–772.

Mantelero, A., Esposito, S., 2021. An Evidence-Based Methodology for Human Rights Impact Assessment (HRIA) in the Development of AI Data-Intensive Systems. Computer Law & Security Review 41, 1–35.

Mantovani, E., Böröcz, I., Hert, P. de, 2020. Conducting research with school children and data in line with "ethical principles" lawyers at work in the ethics management of the H2020 mathisis project. Computer Law & Security Review 38, 105451.

Mao, Y., Shi-Kupfer, K., 2021. Online public discourse on artificial intelligence and ethics in China: context, content, and implications. AI & SOCIETY, 1–17.

Marella, V., Upreti, B., Merikivi, J., Tuunainen, V.K., 2020. Understanding the creation of trust in cryptocurrencies: the case of Bitcoin. Electronic Markets 30 (2), 259–271.

Marikyan, D., Papagiannidis, S., Alamanos, E., 2021. "Smart Home Sweet Smart Home". International Journal of E-Business Research 17 (2), 1–23.

Mark, R., 2019. Ethics of Public Use of AI and Big Data. The ORBIT Journal 2 (2), 1–33.

Martin, D., 2017. Who Should Decide How Machines Make Morally Laden Decisions? Sci Eng Ethics 23 (4), 951–967.

Martin, K., 2019. Ethical Implications and Accountability of Algorithms. J Bus Ethics 160 (4), 835–850.

Martin, K.D., Murphy, P.E., 2017. The role of data privacy in marketing. Journal of the Academy of Marketing Science 45 (2), 135–155.

Martinho, A., Kroesen, M., Chorus, C., 2021a. A healthy debate: Exploring the views of medical doctors on the ethics of artificial intelligence. Artificial intelligence in medicine 121, 102190.

Martinho, A., Kroesen, M., Chorus, C., 2021b. Computer Says I Don't Know: An Empirical Approach to Capture Moral Uncertainty in Artificial Intelligence. Minds & Machines 31, 215–237.

Martins-Turner, K., Grahle, A., Nagel, K., Göhlich, D., 2020. Electrification of urban freight transport—a case study of the food retailing industry. Procedia Computer Science 170, 757–763.

May So, W.C., Danny Wong, T.N., Sculli, D., 2005. Factors affecting intentions to purchase via the internet. Industrial Management & Data Systems 105 (9), 1225–1244.

Mayer, H.O., 2013. Interview und schriftliche Befragung: Grundlagen und Methoden empirischer Sozialforschung, 6th ed. Oldenbourg Wissenschaftsverlag GmbH, München.

Mayring, P., 1994. Qualitative Inhaltsanalyse. UVK Univ.-Verl. Konstanz 14, 159–175.

Mayring, P., Fenzl, T., 2019. Qualitative Inhaltsanalyse, in: Baur, N., Blasius, J. (Eds.), Handbuch Methoden der empirischen Sozialforschung. Springer VS, Wiesbaden, 633–648.

McAleenan, P., 2020. Moral responsibility and action in the use of artificial intelligence in construction. Proceedings of the Institution of Civil Engineers—Management, Procurement and Law 173 (4), 166–174.

McDougall, R.J., 2019. Computer knows best? The need for value-flexibility in medical AI. Journal of medical ethics 45 (3), 156–160.

McKernan, L.C., Clayton, E.W., Walsh, C.G., 2018. Protecting Life While Preserving Liberty: Ethical Recommendations for Suicide Prevention With Artificial Intelligence. Frontiers in psychiatry 9, 650.

McKinsey & Company, 2021. Returning to order: Improving returns management for apparel companies <https://www.mckinsey.com/industries/retail/our-insights/returning-to-order-improving-returns-management-for-apparel-companies/> (accessed 5 October 2024).

McKinsey & Company, 2024. Technology Trends Outlook 2024. McKinsey Digital 07.

McKinsey & Company and Global Fashion Agenda, 2020. Fashion on Climate: How the Fashion Industry can urgently act to reduce its greenhouse gas emissions. McKinsey & Company and Global Fashion Agenda.

McKnight, D.H., Choudhury, V., 2006. Distrust and trust in B2C e-commerce: do they differ? Proceedings of the 8th international conference on Electronic commerce The new e-commerce: innovations for conquering current barriers, obstacles and limitations to conducting successful business on the internet, 482–491.

McKnight, D.H., Choudhury, V., Kacmar, C., 2002. Developing and Validating Trust Measures for e-Commerce: An Integrative Typology. Information Systems Research 13 (3), 334–359.

McKnight, D.H., Lankton, N.K., Nicolaou, A., Price, J., 2017. Distinguishing the effects of B2B information quality, system quality, and service outcome quality on trust and distrust. The Journal of Strategic Information Systems 26 (2), 118–141.

McLeod, F.N., Cherrett, T.J., Bektas, T., Allen, J., Martinez-Sykora, A., Lamas-Fernandez, C., …, Wise, S., 2020. Quantifying environmental and financial benefits of using porters and cycle couriers for last-mile parcel delivery. Transportation Research Part D: Transport and Environment 82, 102311.

McManus, R.M., Rutchick, A.M., 2019. Autonomous Vehicles and the Attribution of Moral Responsibility. Social Psychological and Personality Science 10 (3), 345–352.

McStay, A., 2020. Emotional AI, soft biometrics and the surveillance of emotional life: An unusual consensus on privacy. Big Data & Society 7 (1), 205395172090438.

McStay, A., Rosner, G., 2021. Emotional artificial intelligence in children's toys and devices: Ethics, governance and practical remedies. Big Data & Society 8 (1), 205395172199487.

Meese, R.A., Rogoff, K., 1983. Empirical exchange rate models of the seventies. Journal of International Economics 14 (1–2), 3–24.

Mehta, P., Jebarajakirthy, C., Maseeh, H.I., Anubha, A., Saha, R., Dhanda, K., 2022. Artificial intelligence in marketing: A meta-analytic review. Psychology & Marketing 39 (11), 2013–2038.

Meitinger, T., 2021. Studie: Sechs Anbieter dominieren den deutschen Paketmarkt <https://logistik-heute.de/news/studie-sechs-anbieter-dominieren-den-deutschen-paketmarkt-32737.html> (accessed 20 October 2023).

Mello Bandeira, R.A.d., Goes, G.V., Goncalves, D.N.S., Marcio de Almeida, D.A., Oliveira, C.M.d., 2019. Electric vehicles in the last mile of urban freight transportation: A sustainability assessment of postal deliveries in Rio de Janeiro-Brazil. Transportation Research Part D: Transport and Environment 67, 491–502.

Mendez, G., 2024. Top fashion ecommerce trends for 2025—The future of fashion retail <https://www.channelengine.com/en/blog/fashion-ecommerce-trends> (accessed 18 December 2024).

Mendoza-Tello, J.C., Mora, H., Pujol-López, F.A., Lytras, M.D., 2019. Disruptive innovation of cryptocurrencies in consumer acceptance and trust. Information Systems and e-Business Management 17 (2–4), 195–222.

Merritt, S.M., Heimbaugh, H., LaChapell, J., Lee, D., 2013. I trust it, but I don't know why: Effects of implicit attitudes toward automation on trust in an automated system. Human Factors 55 (3), 520–534.

Mersky, A.C., Sprei, F., Samaras, C., Qian, Z., 2016. Effectiveness of incentives on electric vehicle adoption in Norway. Transportation Research Part D: Transport and Environment 46 (3), 56–68.

Meyer, C., Schwager, A., 2007. Understanding customer experience. Harvard Business Review 85 (2), 116.

Miell, S., Gill, S., Vazquez, D., 2018. Enabling the digital fashion consumer through fit and sizing technology. Journal of Global Fashion Marketing 9 (1), 9–23.

Miernicki, M., Ng, I., 2021. Artificial intelligence and moral rights. AI & Soc 36 (1), 319–329.

Miguel, I. de, Sanz, B., Lazcoz, G., 2020. Machine learning in the EU health care context: exploring the ethical, legal and social issues. Information, Communication & Society 23 (8), 1139–1153.

Mikalef, P., Conboy, K., Lundström, J.E., Popovič, A., 2022. Thinking responsibly about responsible AI and 'the dark side' of AI. European Journal of Information Systems 31 (3), 257–268.

Milano, S., Taddeo, M., Floridi, L., 2020. Recommender systems and their ethical challenges. AI & Soc 35 (4), 957–967.

Milano, S., Taddeo, M., Floridi, L., 2021. Ethical aspects of multi-stakeholder recommendation systems. The Information Society 37 (1), 35–45.

Miles, M.B., Hubermann, A.M., 1994. Qualitative data analysis: An expanded sourcebook. sage.

Miller, S., 2019. Machine Learning, Ethics and Law. 1 23.

Milossi, M., Alexandropoulou-Egyptiadou, E., Psannis, K.E., 2021. AI Ethics: Algorithmic Determinism or Self-Determination? The GPDR Approach. IEEE Access 9, 58455–58466.

Mirbabaie, M., Marx, J., Germies, J., 2021. Conscious Commerce—Digital Nudging and Sustainable E-commerce Purchase Decisions. Proceedings of the 2021 Australasian Conference on Information Systems, Sydney, Australia, 14.

Mirsch, T., Lehrer, C., Jung, R., 2017. Digital nudging: Altering user behavior in digital environments. Proceedings of the 2017 Internationale Tagung Wirtschaftsinformatik 13, 634–648.

Misselhorn, C., 2018. Artificial Morality. Concepts, Issues and Challenges. Soc 55 (2), 161–169.

Mittelstadt, B., 2019. Principles alone cannot guarantee ethical AI. Nat Mach Intell 1 (11), 501–507.

Miyazaki, A.D., Fernandez, A., 2001. Consumer Perceptions of Privacy and Security Risks for Online Shopping. Journal of Consumer Affairs 35 (1), 27–44.

Mökander, J., Axente, M., 2021. Ethics-based auditing of automated decision-making systems: intervention points and policy implications. AI & Soc, 1–19.

Mökander, J., Floridi, L., 2021. Ethics-Based Auditing to Develop Trustworthy AI. Minds & Machines 31 (2), 323–327.

Momani, A.M., 2020. The Unified Theory of Acceptance and Use of Technology: A New Approach in Technology Acceptance. International Journal of Sociotechnology and Knowledge Development 12 (3), 79–98.

Momani, A.M., Jamous, M.M., 2017. The Evolution of Technology Acceptance Theories. International Journal of Contemporary Computer Research 1 (1), 51–58.

Montemayor, C., Halpern, J., Fairweather, A., 2021. In principle obstacles for empathic AI: why we can't replace human empathy in healthcare. AI & Soc, 1–7.

Moody, G.D., Lowry, P.B., Galletta, D.F., 2017. It's complicated: explaining the relationship between trust, distrust, and ambivalence in online transaction relationships using polynomial regression analysis and response surface analysis. European Journal of Information Systems 26 (4), 379–413.

Moon, Y., Nass, C., 1998. Are computers scapegoats? Attributions of responsibility in human–computer interaction. International Journal of Human-Computer Studies 49 (1), 79–94.

Mörch, C.-M., Atsu, S., Cai, W., Li, X., Madathil, S.A., Liu, X., Mai, V., Tamimi, F., Dilhac, M.A., Ducret, M., 2021. Artificial Intelligence and Ethics in Dentistry: A Scoping Review. Journal of Dental Research 100 (13), 1452–1460.

Mörch, C.-M., Gupta, A., Mishara, B.L., 2020. Canada protocol: An ethical checklist for the use of artificial Intelligence in suicide prevention and mental health. Artificial intelligence in medicine 108, 101934.

Morganti, E., Browne, M., 2018. Technical and operational obstacles to the adoption of electric vans in France and the UK: An operator perspective. Transport Policy 63 (3), 90–97.

Mori, M., MacDorman, K.F., Kageki, N., 2012. The Uncanny Valley [from the field]. IEEE Robotics & Automation Magazine 19 (2), 98–100.

Morley, J., Elhalal, A., Garcia, F., Kinsey, L., Mökander, J., Floridi, L., 2021a. Ethics as a Service: A Pragmatic Operationalisation of AI Ethics. Minds & Machines 31 (2), 239–256.

Morley, J., Kinsey, L., Elhalal, A., Garcia, F., Ziosi, M., Floridi, L., 2021b. Operationalising AI ethics: barriers, enablers and next steps. AI & SOCIETY 38, 411–423.

Morley, J., Machado, C.C.V., Burr, C., Cowls, J., Joshi, I., Taddeo, M., Floridi, L., 2020. The ethics of AI in health care: A mapping review. Social science & Medicine 260, 113172.

Mosakas, K., 2021. On the moral status of social robots: considering the consciousness criterion. AI & Soc 36 (2), 429–443.

Mowbray, M., 2021. Moral Status for Malware! The Difficulty of Defining Advanced Artificial Intelligence. Camb Q Healthc Ethics 30 (3), 517–528.

Munir, S., 2020. Eco-Fashion Adoption in the UAE: Understanding Consumer Barriers and Motivational Factors. Fashion Practice 12 (3), 371–393.

Murali, N.G., Ethiraj, M., Aarthi, S., 2017. IOT Based Interactive Smart Refrigerator. 3rd International Conference on Computers and Management.

Murdock, G., 2018. Media Materialties: For A Moral Economy of Machines. Journal of Communication 68 (2), 359–368.

Myers, M.D., Avison, D., 1997. Qualitative research in information systems. SAGE Publications.

Nakamoto, S., 2008. Bitcoin: A Peer-to-Peer Electronic Cash System <http://bitcoin.org/bitcoin.pdf> (accessed 21 August 2023).

Nallur, V., 2020. Landscape of Machine Implemented Ethics. Sci Eng Ethics 26 (5), 2381–2399.

Natarajan, T., Balasubramanian, S.A., Kasilingam, D.L., 2018. The moderating role of device type and age of users on the intention to use mobile shopping applications. Technology in society 53, 79–90.

Nath, R., Manna, R., 2021. From posthumanism to ethics of artificial intelligence. AI & SOCIETY.

Nath, R., Sahu, V., 2020. The problem of machine ethics in artificial intelligence. AI & Soc 35 (1), 103–111.

Nelms, T.C., Maurer, B., Swartz, L., Mainwaring, S., 2018. Social Payments: Innovation, Trust, Bitcoin, and the Sharing Economy. Theory, Culture & Society 35 (3), 13–33.

Nemitz, P., 2018. Constitutional democracy and technology in the age of artificial intelligence. Philosophical transactions. Series A, Mathematical, physical, and engineering sciences 376 (2133).

Nestler, A., Karessli, N., Hajjar, K., Weffer, R., Shirvany, R., 2021. SizeFlags: Reducing Size and Fit Related Returns in Fashion E-Commerce. Proceedings of the 27th ACM SIGKDD Conference on Knowledge Discovery & Data Mining.

Neumann, J. von, Morgenstern, O., 1947. Theory of games and economic behavior, 2nd rev, 2nd ed. Princeton University Press.

Newman, P., 2020. The Internet of Things 2020: Here's what over 400 IoT decision-makers say about the future of enterprise connectivity and how IoT companies can use it to grow revenue <https://www.businessinsider.com/internet-of-things-report> (accessed 2 February 2022).

Nguyen, H.M., Khoa, B.T., 2019. The Relationship between the Perceived Mental Benefits, Online Trust, and Personal Information Disclosure in Online Shopping. The Journal of Asian Finance, Economics and Business 6 (4), 261–270.

Nguyen, T.D., Huynh, P.A., 2018. The Roles of Perceived Risk and Trust on E–Payment Adoption. Econometrics for Financial Applications, 926–940.

Nigam, A., Pasricha, R., Singh, T., Churi, P., 2021. A Systematic Review on AI-based Proctoring Systems: Past, Present and Future. Educ Inf Technol 26 (5), 1–25.

Nikolaou, I.E., Evangelinos, K.I., Allan, S., 2013. A reverse logistics social responsibility evaluation framework based on the triple bottom line approach. Journal of Cleaner Production 56, 173–184.

Noel, L., Rubens, G.Z. de, Kester, J., Sovacool, B.K., 2020. Understanding the socio-technical nexus of Nordic electric vehicle (EV) barriers: A qualitative discussion of range, price, charging and knowledge. Energy Policy 138, 111292.

Noothigattu, R., Bouneffouf, D., Mattei, N., Chandra, R., Madan, P., Varshney, K.R., Campbell, M., Singh, M., Rossi, F., 2019. Teaching AI agents ethical values using reinforcement learning and policy orchestration. IBM J. Res. & Dev. 63 (4/5), 2:1–2:9.

Noppers, E.H., Keizer, K., Bolderdijk, J.W., Steg, L., 2014. The adoption of sustainable innovations: Driven by symbolic and environmental motives. Global Environmental Change 25, 52–62.

Norberg, P.A., Horne, D.R., Horne, D.A., 2007. The Privacy Paradox: Personal Information Disclosure Intentions versus Behaviors. Journal of Consumer Affairs 41 (1), 100–126.

Novak, T.P., 2020. A Generalized Framework for Moral Dilemmas Involving Autonomous Vehicles: A Commentary on Gill. Journal of Consumer Research 47 (2), 292–300.

Ofori, K.S., Boateng, H., Okoe, A.F., Gvozdanovic, I., 2017. Examining customers' continuance intentions towards internet banking usage. Marketing Intelligence & Planning 35 (6), 756–773.

ÓhÉigeartaigh, S.S., Whittlestone, J., Liu, Y., Zeng, Y., Liu, Z., 2020. Overcoming Barriers to Cross-cultural Cooperation in AI Ethics and Governance. Philos. Technol. 33 (4), 571–593.

Okada, E.M., 2005. Justification Effects on Consumer Choice of Hedonic and Utilitarian Goods. Journal of Marketing Research 42 (1), 43–53.

Oreg, S., 2003. Resistance to change: developing an individual differences measure. The Journal of applied psychology 88 (4), 680–693.

Orr, W., Davis, J.L., 2020. Attributions of ethical responsibility by Artificial Intelligence practitioners. Information, Communication & Society 23 (5), 719–735.

Osório, A., Pinto, A., 2019. Information, uncertainty and the manipulability of artificial intelligence autonomous vehicles systems. International Journal of Human-Computer Studies 130, 40–46.

Pagallo, U., 2017. When Morals Ain't Enough: Robots, Ethics, and the Rules of the Law. Minds & Machines 27 (4), 625–638.

Pan, X., Sinha, P., Chen, X., 2021. Corporate social responsibility and eco-innovation: The triple bottom line perspective. Corporate Social Responsibility and Environmental Management 28 (1).

Pandey, S., Chawla, D., 2018. Online customer experience (OCE) in clothing e-retail: Exploring OCE dimensions and their impact on satisfaction and loyalty—Does gender matter? International Journal of Retail & Distribution Management 46 (3).

Parasuraman, A., 2000. Technology Readiness Index (TRI) a multiple-item scale to measure readiness to embrace new technologies. Journal of Service Research 2 (4), 307–320.

Park, S.H., Kim, Y.-H., Lee, J.Y., Yoo, S., Kim, C.J., 2019. Ethical challenges regarding artificial intelligence in medicine from the perspective of scientific editing and peer review. Sci Ed 6 (2), 91–98.

Pavlou, P.A., 2003. Consumer Acceptance of Electronic Commerce: Integrating Trust and Risk with the Technology Acceptance Model. International Journal of Electronic Commerce 7 (3), 101–134.

Pavlou, P.A., Gefen, D., 2004. Building Effective Online Marketplaces with Institution-Based Trust. Information Systems Research 15 (1), 37–59.

PayPal, 2020a. PayPal Launches New Service Enabling Users to Buy, Hold and Sell Cryptocurrency <https://www.youtube.com/watch?v=llPDNu7fTtk> (accessed 21 August 2023).

PayPal, 2020b. Third Quarter 2020 Results <https://s1.q4cdn.com/633035571/files/doc_financials/2020/q3/Q3-20-PayPal-Earnings-Release.pdf> (accessed 13 April 2021).

PayPal, 2021a. How PayPal works <https://www.paypal.com/us/webapps/mpp/how-paypal-works> (accessed 21 April 2021).

PayPal, 2021b. PayPal Investor Day 2021 Fireside Chat: Our Strategies for Increasing Consumer Engagement <https://www.youtube.com/watch?v=tWz7EKItKbM> (accessed 21 August 2023).

Peters, D., Vold, K., Robinson, D., Calvo, R.A., 2020. Responsible AI—Two Frameworks for Ethical Design Practice. IEEE Trans. Technol. Soc. 1 (1), 34–47.

Petersen, K., Vakkalanka, S., Kuzniarz, L., 2015. Guidelines for conducting systematic mapping studies in software engineering: An update. Information and Software Technology 64, 1–18.

Peterson, C., Stunkard, A.J., 1992. Cognates of personal control: Locus of control, self-efficacy, and explanatory style. Applied and Preventive Psychology 1 (2), 111–117.

Pillai, V.S., Matus, K.J.M., 2021. Towards a responsible integration of artificial intelligence technology in the construction sector. Science and Public Policy 47 (5), 689–704.

Pinka, R., 2021. Synthetic Deliberation: Can Emulated Imagination Enhance Machine Ethics? Minds & Machines 31 (1), 121–136.

Podschwadek, F., 2017. Do androids dream of normative endorsement? On the fallibility of artificial moral agents. Artif Intell Law 25 (3), 325–339.

Popa, E.O., van Hilten, M., Oosterkamp, E., Bogaardt, M.-J., 2021. The use of digital twins in healthcare: socio-ethical benefits and socio-ethical risks. Life Sci Soc Policy 17 (1), 6.

Prestes, E., Houghtaling, M.A., Goncalves, P.J., Fabiano, N., Ulgen, O., Fiorini, S.R., Murahwi, Z., Olszewska, J.I., Haidegger, T., 2021. The First Global Ontological Standard for Ethically Driven Robotics and Automation Systems [Standards]. IEEE Robotics & Automation Magazine 28 (4), 120–124.

PricewaterhouseCoopers, 2021. Global Consumer Insights Survey 2021—Pulse 3. PricewaterhouseCoopers GmbH Wirtschaftsprüfungsgesellschaft, Frankfurt am Main, Germany, 1–16.

PricewaterhouseCoopers, 2024a. PwC's 2023 Emerging Technology Survey: The state of emerging tech today—and four practices that grow the top and bottom lines <https://www.pwc.com/us/en/tech-effect/emerging-tech/emtech-survey.html> (accessed 1 October 2024).

PricewaterhouseCoopers, 2024b. The new Essential Eight technologies: what you need to know <https://www.pwc.com/us/en/tech-effect/emerging-tech/essential-eight-technologies.html> (accessed 1 October 2024).

Puca, R.M., Schüler, J., 2017. Motivation, in: Müsseler, J., Rieger, M. (Eds.), Allgemeine Psychologie, 3. Auflage ed. Springer, Berlin, Heidelberg, 223–246.

Puntoni, S., Walker Reczek, R., Giesler, M., Botti, S., 2021. Consumers and Artificial Intelligence: An Experimental Perspective. Journal of Marketing 85 (1), 131–151.
Purvis, B., Mao, Y., Robinson, D., 2019. Three pillars of sustainability: in search of conceptual origins. Sustainability Science 14 (3), 681–695.
Quak, H., Nesterova, N., Van Rooijen, T., 2016. Possibilities and barriers for using electric-powered vehicles in city logistics practice. Transportation Research Procedia 12, 157–169.
Raab, C.D., 2020. Information privacy, impact assessment, and the place of ethics. Computer Law & Security Review 37, 105404.
Radu, R., 2021. Steering the governance of artificial intelligence: national strategies in perspective. Policy and Society 40 (2), 178–193.
Ram, S., Sheth, J.N., 1989. Consumer Resistance to Innovations: The Marketing Problem and its solutions. Journal of Consumer Marketing 6 (2), 5–14.
Reddy, S., Allan, S., Coghlan, S., Cooper, P., 2020. A governance model for the application of AI in health care. Journal of the American Medical Informatics Association : JAMIA 27 (3), 491–497.
Reed, A., Forehand, M.R., Puntoni, S., Warlop, L., 2012. Identity-based consumer behavior. International Journal of Research in Marketing 29 (4), 310–321.
Reed, N., Leiman, T., Palade, P., Martens, M., Kester, L., 2021. Ethics of automated vehicles: breaking traffic rules for road safety. Ethics and Information Technology 23 (4), 777–789.
Rees, C., 2023. The Future of Returns in Retail <https://www.deloitte.com/uk/en/Industries/consumer/blogs/2023/the-future-of-returns-in-retail.html> (accessed 3 October 2024).
Rehman, M.H.u., Salah, K., Damiani, E., Svetinovic, D., 2020. Trust in Blockchain Cryptocurrency Ecosystem. IEEE Transactions on Engineering Management 67 (4), 1196–1212.
Reichheld, A., Peto, J., Ritthaler, C., 2023. Research: Consumers' Sustainability Demands Are Rising <https://hbr.org/2023/09/research-consumers-sustainability-demands-are-rising> (accessed 5 February 2024).
Reid, F., Harrigan, M., 2013. An Analysis of Anonymity in the Bitcoin System, in: Altshuler, Y., Elovici, Y., Cremers, A.B., Aharony, N., Pentland, A. (Eds.), Security and Privacy in Social Networks. Springer, New York, NY, 197–223.
Reinartz, W., Wiegand, N., Imschloss, M., 2019. The impact of digital transformation on the retailing value chain. International Journal of Research in Marketing 36 (3), 350–366.
Reiss, M.J., 2021. The use of AI in education: Practicalities and ethical considerations. London Review of Education 19 (1).
Rejeb, A., Rejeb, K., Treiblmaier, H., 2023. How augmented reality impacts retail marketing: a state-of-the-art review from a consumer perspective. Journal of Strategic Marketing 31 (3), 718–748.
Rességuier, A., Rodrigues, R., 2020. AI ethics should not remain toothless! A call to bring back the teeth of ethics. Big Data & Society 7 (2), 205395172094254.
Reynolds, K., Voell, Z., 2020. Almost 20% of PayPal Users Have Used App to Trade Bitcoin, Mizuho Says <https://www.coindesk.com/markets/2020/12/01/almost-20-of-paypal-users-have-used-app-to-trade-bitcoin-mizuho-says/> (accessed 21 August 2023).
Rezvani, Z., Jansson, J., Bengtsson, M., 2018. Consumer motivations for sustainable consumption: The interaction of gain, normative and hedonic motivations on electric vehicle adoption. Business Strategy and the Environment 27 (8), 1272–1283.

Rezvani, Z., Jansson, J., Bodin, J., 2015. Advances in consumer electric vehicle adoption research: A review and research agenda. Transportation Research Part D: Transport and Environment 34, 122–136.

Rhim, J., Lee, J.-H., Chen, M., Lim, A., 2021. A Deeper Look at Autonomous Vehicle Ethics: An Integrative Ethical Decision-Making Framework to Explain Moral Pluralism. Front. Robot. AI 8, 632394.

Ritchie, H., 2020. Cars, planes, trains: where do CO2 emissions from transport come from? <https://ourworldindata.org/co2-emissions-from-transport>.

Robbins, S., 2019. A Misdirected Principle with a Catch: Explicability for AI. Minds & Machines 29 (4), 495–514.

Robbins, S., 2020. AI and the path to envelopment: knowledge as a first step towards the responsible regulation and use of AI-powered machines. AI & Soc 35 (2), 391–400.

Roberts, H., Cowls, J., Morley, J., Taddeo, M., Wang, V., Floridi, L., 2021. The Chinese approach to artificial intelligence: an analysis of policy, ethics, and regulation. AI & Soc 36 (1), 59–77.

Roberts, J., Grassi, A., 2021. A review of studies on virtual layout and atmospherics—potential applications to the fashion industry. The International Review of Retail, Distribution and Consumer Research 31 (4), 432–456.

Robra-Bissantz, S., Lattemann, C., 2019. 7 Rules of Attraction—Mit kundenorientierten Diensten erfolgreich in der Digitalen Transformation, in: , Digital Customer Experience. Springer Vieweg, Wiesbaden, 3–22.

Roca-Puigros, M., Marmy, C., Wäger, P., Müller, D.B. Modeling the transition toward a zero emission car fleet: Integrating electrification, shared mobility, and automation. Transportation Research Part D: Transport and Environment 115, 103576.

Rochel, J., Evéquoz, F., 2021. Getting into the engine room: a blueprint to investigate the shadowy steps of AI ethics. AI & Soc 36 (2), 609–622.

Roff, H.M., 2019. Artificial Intelligence: Power to the People. Ethics int. aff. 33 (02), 127–140.

Rogers, E.M., 1995. Diffusion of Innovations. Free Press, New York.

Ryan, M., 2019. Ethics of Using AI and Big Data in Agriculture: The Case of a Large Agriculture Multinational. The ORBIT Journal 2 (2), 1–27.

Ryan, M., 2020. In AI We Trust: Ethics, Artificial Intelligence, and Reliability. Sci Eng Ethics 26 (5), 2749–2767.

Ryan, M., Anya, G., 2019. Ethics of Using Smart City AI and Big Data: The Case of Four Large European Cities. The ORBIT Journal 2 (2), 1–36.

Ryan, M., Stahl, B.C., 2021. Artificial intelligence ethics guidelines for developers and users: clarifying their content and normative implications. JICES 19 (1), 61–86.

Sachdeva, I., Goel, S., 2015. Retail Store Environment and Customer Experience: A Paradigm. Journal of Fashion Marketing and Management 19 (3), 290–298.

Sadiq, M., Adil, M., Paul, J., 2021. An innovation resistance theory perspective on purchase of eco-friendly cosmetics. Journal of Retailing and Consumer Services 59, 102369.

Sætra, H.S., 2020. A shallow defence of a technocracy of artificial intelligence: Examining the political harms of algorithmic governance in the domain of government. Technology in society 62, 101283.

Saheb, T., Saheb, T., Carpenter, D.O., 2021. Mapping research strands of ethics of artificial intelligence in healthcare: A bibliometric and content analysis. Computers in biology and medicine 135.

Salem, F.A., Moreno, U.F., Lamnabhi-Lagarrigue, F., 2020. A control approach to address ethical issues on social (robotic) networks. IFAC-PapersOnLine 53 (2), 17016–17022.

Sand, M., Durán, J.M., Jongsma, K.R., 2021. Responsibility beyond design: Physicians' requirements for ethical medical AI. Bioethics.

Sandin, G., Roos, S., Spak, B., Zamani, B., Peters, G., 2019. The impact of Swedish clothing consumption <http://mistrafuturefashion.com/impact-of-swedish-clothing-consumption/> (accessed 25 October 2024).

Sangari, M.S., Mashatan, A., 2024. What is driving consumer resistance to crypto-payment? A multianalytical investigation. Psychology & Marketing 41 (3), 575–591.

Sas, C., Khairuddin, I.E., 2015. Exploring Trust in Bitcoin Technology. Proceedings of the Annual Meeting of the Australian Special Interest Group for Computer Human Interaction, 338–342.

Sas, C., Khairuddin, I.E., 2017. Design for Trust: An exploration of the challenges and opportunities of bitcoin users. Proceedings of the 2017 CHI Conference on Human Factors in Computing Systems, 6499–6510.

Savitz, A., 2013. The Tripple Bottom Line: How Today's Best-Run Campanies are Achieving Economic, Social and Environmental Success- and how you can too. John Wiley & Sons.

Schacker, M., Stanoevska-Slabeva, K., 2021. Understanding the Impact of Sustainability and CSR Information in D2C Online Shops on Consumer Attitudes and Behavior—A Literature Review. Proceedings of the 2021 Hawaii International Conference on System Sciences, Hawaii, USA.

Schanke, S., Burtch, G., Ray, G., 2021. Estimating the Impact of "Humanizing" Customer Service Chatbots. Information Systems Research 32 (3), 736–751.

Schendera, C.F., 2011. Datenqualität mit SPSS. Oldenbourg Verlag.

Scheutz, M., 2017. The Case for Explicit Ethical Agents. AIMag 38 (4), 57–64.

Schiff, D., 2021. Education for AI, not AI for Education: The Role of Education and Ethics in National AI Policy Strategies. International Journal of Artificial Intelligence in Education, 1–37.

Schmidt, A.T., Engelen, B., 2020. The ethics of nudging: An overview. Philosophy Compass 15 (4).

Schomakers, E.-M., Klatte, M., Lotz, V., Biermann, H., Kober, F., Ziefle, M., 2022. Analysis of the potential of a new concept for urban last-mile delivery: Ducktrain. 2590–1982 14, 100579.

Schönberger, D., 2019. Artificial intelligence in healthcare: a critical analysis of the legal and ethical implications. International Journal of Law and Information Technology 27 (2), 171–203.

Schrader, D.E., Ghosh, D., 2018. Proactively Protecting Against the Singularity: Ethical Decision Making in AI. IEEE Secur. Privacy 16 (3), 56–63.

Schrills, T., Zoubir, M., Stahl, J., Drozniak, K., Franke, T., 2020. Good Boy Here or Bad Boy Far Away? Effects of Digital Nudging on Booking Decisions in Car Sharing. Proceedings of the 2020 International Conference on Human-Computer Interaction, Copenhagen, Denmark, 398–411.

Schuhwerk, M.E., Lefkoff-Hagius, R., 1995. Green or Non-Green? Does Type of Appeal Matter When Advertising a Green Product? Journal of Advertising 24 (2), 45–54.

Schulz, S.A., Flanigan, R.L., 2016. Developing competitive advantage using the triple bottom line: A conceptual framework. Journal of Business & Industrial Marketing 31 (4).

Sebastián, M.Á., Rudy-Hiller, F., 2021. First-person representations and responsible agency in AI. Synthese, 1–19.

Seeber, I., Bittner, E., Briggs, R.O., Vreede, G.-J. de, Vreede, T. de, Druckenmiller, D., Maier, R., Merz, A.B., Oeste-Reiß, S., Randrup, N., Schwabe, G., Söllner, M., 2018. Machines as Teammates: A Collaboration Research Agenda. Proceedings of the 51st Hawaii International Conference on System Sciences.

Segun, S.T., 2021a. Critically engaging the ethics of AI for a global audience. Ethics Inf Technol 23 (2), 99–105.

Segun, S.T., 2021b. From machine ethics to computational ethics. AI & Soc 36 (1), 263–276.

Seidenschwarz, H., Deichner, N., Stahl, E., Wittmann, G., 2020. Erfolgsfaktor Payment—Der Einfluss der Zahlungsverfahren auf den Umsatz <https://ibi.de/veroeffentlichungen/ErfolgsfaktorPayment2020> (accessed 21 August 2023).

Sekiguchi, K., Hori, K., 2020. Organic and dynamic tool for use with knowledge base of AI ethics for promoting engineers' practice of ethical AI design. AI & Soc 35 (1), 51–71.

Sequeiros, H., Oliveira, T., Thomas, M.A., 2022. The Impact of IoT Smart Home Services on Psychological Well-Being. Information Systems Frontiers 24 (3), 1009–1026.

SFR-IA Group, CERF, 2018. Artificial intelligence and medical imaging 2018: French Radiology Community white paper. Diagnostic and Interventional Imaging 99 (11), 727–742.

Shank, D.B., DeSanti, A., 2018. Attributions of morality and mind to artificial intelligence after real-world moral violations. Computers in Human Behavior 86, 401–411.

Shapiro, S., MacInnis, D.J., 2002. Understanding Program-Induced Mood Effects: Decoupling Arousal from Valence. Journal of Advertising 31 (4), 15–26.

Sharma, S., Singh, G., Gaur, L., Afaq, A., 2022. Exploring customer adoption of autonomous shopping systems. Telematics and Informatics 73, 101861.

Shaw, J.A., Sethi, N., Block, B.L., 2021. Five things every clinician should know about AI ethics in intensive care. Intensive care medicine 47 (2), 157–159.

Shehu, E., Papies, D., Neslin, S.A., 2020. Free Shipping Promotions and Product Returns. Journal of Marketing Research 57 (4), 640–658.

Sheinin, D.A., Varki, S., Ashley, C., 2011. The Differential Effect of Ad Novelty and Message Usefulness on Brand Judgments. Journal of Advertising 40 (3), 5–18.

Shell, 2021. The Energy Transformation Scenarios <www.shell.com/transformationscenarios> (accessed 6 July 2023).

Shen, B., Li, Q., Dong, C., Perry, P., 2017. Sustainability Issues in Textile and Apparel Supply Chains. Sustainability 9 (9), 1592.

Shevlin, H., 2021. How Could We Know When a Robot was a Moral Patient? Camb Q Healthc Ethics 30 (3), 459–471.

Shneiderman, B., 2020. Bridging the Gap Between Ethics and Practice. ACM Trans. Interact. Intell. Syst. 10 (4), 1–31.

Siegel, J., Pappas, G., 2021. Morals, ethics, and the technology capabilities and limitations of automated and self-driving vehicles. AI & SOCIETY.

Sierzchula, W., 2014. Factors influencing fleet manager adoption of electric vehicles. Transportation Research Part D: Transport and Environment 31 (3), 126–134.

Silva, V., Amaral, A., Fontes, T., 2023. Sustainable Urban Last-Mile Logistics: A Systematic Literature Review. Sustainability 15 (3), 2285.

Simon, H.A., 1995. Artificial intelligence: an empirical science. Artificial Intelligence 77 (1), 95–127.

Simshaw, D., 2018. Ethical issues in robo-lawyering: The need for guidance on developing and using artificial intelligence in the practice of law. Hastings Law Journal 70 (1), 173–214.

Siragusa, C., Tumino, A., Mangiaracina, R., Perego, A., 2022. Electric vehicles performing last-mile delivery in B2C e-commerce: An economic and environmental assessment. International Journal of Sustainable Transportation 16 (1), 22–33.

Smith, M., Miller, S., 2021. The ethical application of biometric facial recognition technology. AI & Soc, 1–9.

Smith, N., Vickers, D., 2021. Statistically responsible artificial intelligences. Ethics and Information Technology, 1–11.

Sood, K., 2018. The Ultimate Black Box: The Thorny Issue of Programming Moral Standards in Machines [Industry View]. IEEE Technol. Soc. Mag. 37 (2), 27–29.

Sparkes, M., 2021. Will PayPal' adoption of bitcoin make cryptocurrency more mainstream? <https://www.newscientist.com/article/2288039-will-paypals-adoption-of-bitcoin-make-cryptocurrency-more-mainstream> (accessed 21 August 2023).

Sparkman, G., Walton, G.M., 2017. Dynamic Norms Promote Sustainable Behavior, Even if It Is Counternormative. Psychological Science 28 (11), 1663–1674.

Sparrow, R., 2021. Why machines cannot be moral. AI & Soc 36 (3), 685–693.

Spiekermann, S., Grossklags, J., Berendt, B., 2001. E-privacy in 2nd generation E-commerce: privacy preferences versus actual behavior. Proceedings of the 3rd ACM conference on Electronic Commerce, 38–47.

Stahl, B.C., 2012. Morality, Ethics, and Reflection: A Categorization of Normative IS Research. Journal of the association for information systems 13 (8), 636–656.

Stahl, B.C., 2022. Responsible innovation ecosystems: Ethical implications of the application of the ecosystem concept to artificial intelligence. International Journal of Information Management 62.

Stahl, B.C., Andreou, A., Brey, P., Hatzakis, T., Kirichenko, A., Macnish, K., Laulhé Shaelou, S., Patel, A., Ryan, M., Wright, D., 2021a. Artificial intelligence for human flourishing—Beyond principles for machine learning. Journal of Business Research 124, 374–388.

Stahl, B.C., Antoniou Josephina, Ryan, M., Macnish, K., Jiya, T., 2021b. Organisational responses to the ethical issues of artificial intelligence. AI & Soc, 1–15.

Statista, 2023. Fashion e-commerce worldwide—statistics & facts.

Steels, L., Lopez de Mantaras, R., 2018. The Barcelona declaration for the proper development and usage of artificial intelligence in Europe. AIC 31 (6), 485–494.

Steg, L., Bolderdijk, J.W., Keizer, K., Perlaviciute, G., 2014. An Integrated Framework for Encouraging Pro-environmental Behaviour: The role of values, situational factors and goals. Journal of Environmental Psychology 38, 104–115.

Steg, L., Vlek, C., 2009. Encouraging pro-environmental behaviour: An integrative review and research agenda. Journal of Environmental Psychology 29 (3), 309–317.

Stewart, C., Wong, S.K.Y., Sung Joseph JY, 2021. Mapping ethico-legal principles for the use of artificial intelligence in gastroenterology. Journal of Gastroenterology and Hepatology 36 (5), 1143–1148.

Stewart, K.J., 2003. Trust Transfer on the World Wide Web. Organization Science 14 (1), 5–17.

Stewart, K.J., 2006. How Hypertext Links Influence Consumer Perceptions to Build and Degrade Trust Online. Journal of Management Information Systems 23 (1), 183–210.

Stix, C., 2021. Actionable Principles for Artificial Intelligence Policy: Three Pathways. Sci Eng Ethics 27 (1), 1–17.

Stokes, F., Amitabha Palmer, 2020. Artificial Intelligence and Robotics in Nursing: Ethics of Caring as a Guide to Dividing Tasks Between AI and Humans. Nursing Philosophy 21 (4), e12306.

Strahilevitz, M., Myers, J.G., 1998. Donations to Charity as Purchase Incentives: How Well They Work May Depend on What You Are Trying to Sell. Journal of Consumer Research 24 (4), 434–446.

Stryja, C., Satzger, G., 2019. Digital nudging to overcome cognitive resistance in innovation adoption decisions. The Service Industries Journal 39 (15–16), 1123–1139.

Sun, F., Ye, R., 2021. Moral Considerations of Artificial Intelligence. Science & Education.

Sun, Y., Fang, S., Hwang, Y., 2019. Investigating Privacy and Information Disclosure Behavior in Social Electronic Commerce. Sustainability 11 (12), 3311.

Sunstein, C.R., 2014. Nudging: A Very Short Guide. Journal of Consumer Policy 37 (4), 583–588.

Sutanto, J., Palme, E., Tan, C.-H., Phang, C.W., 2013. Addressing the Personalization-Privacy Paradox: An Empirical Assessment from a Field Experiment on Smartphone Users. MIS Quarterly 37 (4), 1141–1164.

Swallow, T., 2023. EY shares consumer sustainability and climate sentiments <https://sustainabilitymag.com/sustainability/ey-shares-consumer-sustainability-and-climate-sentiments> (accessed 6 February 2024).

Swanepoel, D., 2021. The possibility of deliberate norm-adherence in AI. Ethics Inf Technol 23 (2), 157–163.

Swarte, T. de, Boufous, O., Escalle, P., 2019. Artificial intelligence, ethics and human values: the cases of military drones and companion robots. Artif Life Robotics 24 (3), 291–296.

Sweeney, J.C., Soutar, G.N., 2001. Consumer perceived value: The development of a multiple item scale. Journal of Retailing 77 (2), 203–220.

Szocik, K., Abylkasymova, R., 2021. Ethical Issues in Police Robots. The Case of Crowd Control Robots in a Pandemic. Journal of Applied Security Research, 1–16.

Taddeo, M., McNeish, D., Blanchard, A., Edgar, E., 2021. Ethical Principles for Artificial Intelligence in National Defence. Philosophy & Technology 34 (4), 1707–1729.

Tajalli, P., 2021. AI ethics and the banality of evil. Ethics Inf Technol, 1–8.

Talke, K., Heidenreich, S., 2014. How to overcome pro-change bias: incorporating passive and active innovation resistance in innovation decision models. Journal of Product Innovation Management 31 (5), 894–907.

Talwar, S., Talwar, M., Kaur, P., Dhir, A., 2020. Consumers' resistance to digital innovations: A systematic review and framework development. Australasian Marketing Journal 28, 286–299.

Tan, W.-K., Tan, C.-H., Teo, H.-H., 2012. Consumer-based decision aid that explains which to buy: Decision confirmation or overconfidence bias? 53 (1), 127–141.

Tandon, A., Sithipolvanichgul, J., Asmi, F., Anwar, M.A., Dhir, A., 2023. Drivers of green apparel consumption: Digging a little deeper into green apparel buying intentions. Business Strategy and the Environment 32 (6), 3997–4012.

Tandon, U., Kiran, R., Sah, A.N., 2016. Customer satisfaction using website functionality, perceived usability and perceived usefulness towards online shopping in India. Information Development 32 (5), 1657–1673.

Tandon, U., Kiran, R., Sah, A.N., 2018. The influence of website functionality, drivers and perceived risk on customer satisfaction in online shopping: an emerging economy case. Information Systems and e-Business Management 16 (1), 57–91.

Tang, Y., Chen, S., Yuan, Z., 2020. The effects of hedonic, gain, and normative motives on sustainable consumption: Multiple mediating evidence from China. Sustainable Development 28 (4), 741–750.

Teubner, T., Hawlitschek, F., Adam, M.T.P., 2019. Reputation Transfer. Business & Information Systems Engineering 61 (2), 229–235.

Thaler, R.H., Sunstein, C.R., 2009. Nudge: Improving decisions about health, wealth, and happiness. Penguin, Munich, Germany.

Thompson, R.L., Higgins, C.A., Howell, J.M., 1991. Personal Computing: Toward a Conceptual Model of Utilization. MIS Quarterly 15 (1), 125.

Tigard, D.W., 2021. Artificial Moral Responsibility: How We Can and Cannot Hold Machines Responsible. Camb Q Healthc Ethics 30 (3), 435–447.

Timmers, P., 2019. Ethics of AI and Cybersecurity When Sovereignty is at Stake. Minds & Machines 29 (4), 635–645.

Todd, P., Benbasat, I., 2000. Inducing compensatory information processing through decision aids that facilitate effort reduction: an experimental assessment. Journal of Behavioral Decision Making 13 (1), 91–106.

Togni, G. de, Erikainen, S., Chan, S., Cunningham-Burley, S., 2021. What makes AI 'intelligent' and 'caring'? Exploring affect and relationality across three sites of intelligence and care. Social science & medicine (1982) 277, 113874.

Tolmeijer, S., Kneer, M., Sarasua, C., Christen, M., Bernstein, A., 2020. Implementations in Machine Ethics: A Survey. ACM Computing Surveys (CSUR) 53 (6), 1–38.

Tomalin, M., Byrne, B., Concannon, S., Saunders, D., Ullmann, S., 2021. The practical ethics of bias reduction in machine translation: why domain adaptation is better than data debiasing. Ethics Inf Technol, 1–15.

Tomlinson, V., 2020. UPS invests in Arrival and orders 10,000 Generation 2 Electric Vehicles <https://arrival.com/news/ups-invests-in-arrival-and-orders-10000-generation-2-electric-vehicles> (accessed 14 April 2022).

Tornatzky, L.G., Klein, K.J., 1982. Innovation characteristics and innovation adoption-implementation: A meta-analysis of findings. IEEE Trans. Eng. Manage. EM-29 (1), 28–45.

Tranfield, D., Denyer, D., Smart, P., 2003. Towards a Methodology for Developing Evidence-Informed Management Knowledge by Means of Systematic Review. British Journal of Management 14 (3), 207–222.

Truby, J., 2020. Governing artificial intelligence to benefit the UN sustainable development goals. Sustainable Development 28 (4), 946–959.

Truby, J., Brown, R., 2021. Human digital thought clones: the Holy Grail of artificial intelligence for big data. Information & Communications Technology Law 30 (2), 140–168.

Tsakalidis, A., Krause, J., Julea, A., Peduzzi, E., Pisoni, E., Thiel, C., 2020. Electric light commercial vehicles: Are they the sleeping giant of electromobility? Transportation Research Part D: Transport and Environment, 102421.

Tsang, A., Schuetze, C.F., 2018. In a First for Germany, Hamburg Bans Diesel Engines <https://www.nytimes.com/2018/05/30/business/energy-environment/hamburg-germany-diesel.html> (accessed 5 September 2023).

Tversky, A., Kahneman, D., 1981. The framing of decisions and the psychology of choice. Science 211 (4481), 453–458.

Um, T., Kim, T., Chung, N., 2020. How does an Intelligence Chatbot Affect Customers Compared with Self-Service Technology for Sustainable Services? Sustainability 12 (12), 5119.

Umbrello, S., 2020. Imaginative value sensitive design: Using moral imagination theory to inform responsible technology design. Science and Engineering Ethics 26 (2), 575–595.

Umbrello, S., Capasso, M., Balistreri, M., Pirni, A., Merenda, F., 2021. Value Sensitive Design to Achieve the UN SDGs with AI: A Case of Elderly Care Robots. Minds and Machines, 1–25.

United Nations, 2023. The 17 Goals <https://sdgs.un.org/goals> (accessed 17 January 2024).

United Nations Conference on Trade and Development secretariat, 2001. E-Commerce and Development Report 2001 <https://unctad.org/system/files/official-document/ecdr2001_en.pdf> (accessed 21 August 2023).

United Nations, Department of Economic and Social Affairs Population Division, 2019. World Urbanization Prospects: The 2018 Revision (ST/ESA/SER.A/420), New York, United Nations.

Vadakkepatt, G.G., Winterich, K.P., Mittal, V., Zinn, W., Beitelspacher, L., Aloysius, J., Ginger, J., Reilman, J., 2021. Sustainable Retailing. Journal of Retailing 97 (1), 62–80.

Vakkuri, V., Kemell, K.-K., Jantunen, M., Halme, E., Abrahamsson, P., 2021. ECCOLA—A method for implementing ethically aligned AI systems. Journal of Systems and Software 182, 111067.

Vakkuri, V., Kemell, K.-K., Kultanen, J., Abrahamsson, P., 2020. The current state of industrial practice in artificial intelligence ethics. IEEE Software 37 (4), 50–57.

Vallejos, E.P., Wortham, R., Miakinkov, E., 2017. When AI goes to war: youth opinion, fictional reality and autonomous weapons. The ORBIT Journal 1 (1), 1–20.

Vallès-Peris, N., Domènech, M., 2021. Caring in the in-between: a proposal to introduce responsible AI and robotics to healthcare. AI & SOCIETY.

Vamplew, P., Dazeley, R., Foale, C., Firmin, S., Mummery, J., 2018. Human-aligned artificial intelligence is a multiobjective problem. Ethics and Information Technology 20 (1), 27–40.

van Berkel, N., Tag, B., Goncalves, J., Hosio, S., 2020. Human-centred artificial intelligence: a contextual morality perspective. Behaviour & Information Technology, 1–17.

van Bruxvoort, X., van Keulen, M., 2021. Framework for Assessing Ethical Aspects of Algorithms and Their Encompassing Socio-Technical System. Applied Sciences 11 (23).

van de Poel, I., 2020. Embedding values in artificial intelligence (AI) systems. Minds and Machines 30 (3), 385–409.

van der Waa, J., Verdult, S., van den Bosch, K., van Diggelen, J., Haije, T., van der Stigchel, B., Cocu, I., 2021. Moral decision making in human-agent teams: Human control and the role of explanations. Frontiers in Robotics and AI 8, 1–20.

van Gestel, L.C., Adriaanse, M.A., Ridder, D.T.D. de, 2021. Do nudges make use of automatic processing? Unraveling the effects of a default nudge under type 1 and type 2 processing. Comprehensive Results in Social Psychology 5 (1–3), 4–24.

Vass, T. de, Shee, H., Miah, S.J., 2021. Iot in supply chain management: a narrative on retail sector sustainability. International Journal of Logistics Research and Applications 24 (6), 605–624.

Venkatesh, Morris, Davis, 2003. User Acceptance of Information Technology: Toward a Unified View. MIS Quarterly 27 (3), 425.

Venkatesh, Thong, Xu, 2012. Consumer Acceptance and Use of Information Technology: Extending the Unified Theory of Acceptance and Use of Technology. MIS Quarterly 36 (1), 157.

Venkatesh, V., 2022. Adoption and use of AI tools: a research agenda grounded in UTAUT. Annals of Operations Research 308 (1), 641–652.

Venkatesh, V., Davis, F.D., 2000. A Theoretical Extension of the Technology Acceptance Model: Four Longitudinal Field Studies. Management Science 46 (2), 186–204.

Verbeek, P.-P., 2008. Morality in design: Design ethics and the morality of technological artifacts. Philosophy and Design, 91–103.

Vesnic-Alujevic, L., Nascimento, S., Pólvora, A., 2020. Societal and ethical impacts of artificial intelligence: Critical notes on European policy frameworks. Telecommunications Policy 44 (6), 101961.

Vidgen, R., Hindle, G., Randolph, I., 2020. Exploring the ethical implications of business analytics with a business ethics canvas. European Journal of Operational Research 281 (3), 491–501.

Virdi, P., Kalro, A.D., Sharma, D., 2020. Online decision aids: the role of decision-making styles and decision-making stages. International Journal of Retail & Distribution Management 48 (6).

Visa Inc., 2021. Visa Q1 2021 Earnings Call Transcript <www.fool.com/earnings/calltranscripts/2021/01/28/visa-v-q1-2021-earnings-call-transcript/> (accessed 13 April 2021).

Völter, F., Urbach, N., Padget, J., 2023. Trusting the trust machine: Evaluating trust signals of blockchain applications. International Journal of Information Management 68, 102429.

Voss, K.E., Spangenberg, E.R., Grohmann, B., 2003. Measuring the Hedonic and Utilitarian Dimensions of Consumer Attitude. Journal of Marketing Research 40 (3), 310–320.

Vrščaj, D., Nyholm, S., Verbong, G.P.J., 2020. Is tomorrow's car appealing today? Ethical issues and user attitudes beyond automation. AI & Soc 35 (4), 1033–1046.

Vugts, A., van den Hoven, M., Vet, E. de, Verweij, M., 2020. How autonomy is understood in discussions on the ethics of nudging. Behavioural Public Policy 4 (1), 108–123.

Wakefield, R., 2013. The influence of user affect in online information disclosure. The Journal of Strategic Information Systems 22 (2), 157–174.

Wakunuma, K., Jiya, T., Aliyu, S., 2020. Socio-ethical implications of using AI in accelerating SDG3 in Least Developed Countries. Journal of Responsible Technology 4, 100006.

Wang, N., Shen, X.-L., Sun, Y., 2013. Transition of electronic word-of-mouth services from web to mobile context: A trust transfer perspective. Decision Support Systems 54 (3), 1394–1403.

Wang, P., 2019. On Defining Artificial Intelligence. Journal of Artificial General Intelligence 10 (2), 1–37.

Wang, T., Duong, T.D., Chen, C.C., 2016. Intention to disclose personal information via mobile applications: A privacy calculus perspective. International Journal of Information Management 36 (4), 531–542.

Wang, W., Benbasat, I., 2009. Interactive decision aids for consumer decision making in e-commerce: The influence of perceived strategy restrictiveness. MIS Quarterly 33 (2), 293–320.

Wang, W., Benbasat, I., 2016. Empirical Assessment of Alternative Designs for Enhancing Different Types of Trusting Beliefs in Online Recommendation Agents. Journal of Management Information Systems 33 (3), 744–775.

Wang, Y.J., Minor, M.S., Wei, J., 2011. Aesthetics and the online shopping environment: Understanding consumer responses. Journal of Retailing 87 (1), 46–58.

Wang, Z., Liu, Y., 2014. Identifying Key Factors Affecting Information Disclosure Intention in Online Shopping. International Journal of Smart Home 8 (4), 47–58.

Wangmo, T., Lipps, M., Kressig, R.W., Ienca, M., 2019. Ethical concerns with the use of intelligent assistive technology: findings from a qualitative study with professional stakeholders. BMC Med Ethics 20 (1), 1–11.

Wareham, C.S., 2021. Artificial intelligence and African conceptions of personhood. Ethics Inf Technol 23 (2), 127–136.

Wasilow, S., Thorpe, J.B., 2019. Artificial Intelligence, Robotics, Ethics, and the Military: A Canadian Perspective. AIMag 40 (1), 37–48.

Wątorek, M., Drożdż, S., Kwapień, J., Minati, L., Oświęcimka, P., Stanuszek, M., 2021. Multiscale characteristics of the emerging global cryptocurrency market. Physics Reports 901, 1–82.

Weber-Guskar, E., 2021. How to feel about emotionalized artificial intelligence? When robot pets, holograms, and chatbots become affective partners. Ethics Inf Technol, 1–10.

Weinmann, M., Schneider, C., vom Brocke, J., 2016. Digital Nudging. Business & Information Systems Engineering 58 (6), 433–436.

Weiss, L., Johar, G.V., 2013. Egocentric Categorization and Product Judgment: Seeing Your Traits in What You Own (and Their Opposite in What You Don't). Journal of Consumer Research 40 (1), 185–201.

Westin, K., Nordlund, A., Jansson, J., Nilsson, J., 2020. Goal Framing as a Tool for Changing People's Car Travel Behavior in Sweden. Sustainability 12 (9), 3695.

Whittlestone, J., Nyrup, R., Alexandrova, A., Cave, S., 2019. The Role and Limits of Principles in AI Ethics, in: Proceedings of the 2019 AAAI/ACM Conference on AI, Ethics, and Society. AIES '19: AAAI/ACM Conference on AI, Ethics, and Society, Honolulu HI USA. 27 01 2019 28 01 2019. Association for Computing Machinery, New York, NY, United States, 195–200.

Wijland, R., Hansen, P., Gardezi, F., 2016. Mobile nudging: Youth engagement with banking apps. Journal of Financial Services Marketing 21 (1), 51–63.

Wikström, M., Hansson, L., Alvfors, P., 2016. Investigating barriers for plug-in electric vehicle deployment in fleets. Transportation Research Part D: Transport and Environment 49, 59–67.

Willems, I., 2022. Warum das Internet of Things KI braucht <https://dmexco.com/de/stories/darum-braucht-das-internet-of-things-kuenstliche-intelligenz/> (accessed 2 February 2022).

Winfield, A.F., Michael, K., Pitt, J., Evers, V., 2019. Machine Ethics: The Design and Governance of Ethical AI and Autonomous Systems [Scanning the Issue]. Proc. IEEE 107 (3), 509–517.

Winfield, A.F.T., Jirotka, M., 2018. Ethical governance is essential to building trust in robotics and artificial intelligence systems. Philosophical Transactions of the Royal Society A: Mathematical, Physical and Engineering Sciences 376 (2133), 1–13.

Wirtz, B.W., Müller, W.M., 2019. An integrated artificial intelligence framework for public management. Public Management Review 21 (7), 1076–1100.

Wohlin, C., Runeson, P., Höst, M., Ohlsson, M.C., 2012. Experimentation in Software Engineering. Springer Berlin Heidelberg, Berlin.

Wolff, A., Gomez-Pilar, J., Nakao, T., Northoff, G., 2019. Interindividual neural differences in moral decision-making are mediated by alpha power and delta/theta phase coherence. Sci Rep 9 (1), 1–13.

Wolff, S., Madlener, R., 2019. Driven by change: Commercial drivers' acceptance and efficiency perceptions of light-duty electric vehicle usage in Germany. Transportation Research Part C: Emerging Technologies 105, 262–282.

Wood, W., Neal, D.T., 2009. The habitual consumer. Journal of Consumer Psychology 19 (4), 579–592.

Wu, F., Lu, C., Zhu, M., Chen, H., Zhu, J., Yu, K., Li, L., Li, M., Chen, Q., Li, X., Cao, X., Wang, Z., Zha, Z., Zhuang, Y., Pan, Y., 2020a. Towards a new generation of artificial intelligence in China. Nat Mach Intell 2 (6), 312–316.

Wu, W., Huang, T., Gong, K., 2020b. Ethical Principles and Governance Technology Development of AI in China. Engineering 6 (3), 302–309.

Wünderlich, N.V., Heinonen, K., Ostrom, A.L., Patrício, L., Sousa, R., Voss, C., Lemmink, J.G., 2015. "Futurizing" smart service: implications for service researchers and managers. Journal of Services Marketing 29 (6/7), 442–447.

Wünderlich, N.V., Wangenheim, F.v., Bitner, M.J., 2013. High Tech and High Touch. Journal of Service Research 16 (1), 3–20.

Xiao, B., Benbasat, I., 2007. E-Commerce Product Recommendation Agents: Use, Characteristics, and Impact. MIS Quarterly 31 (1), 137.

Xue, Y., Zhang, X., Zhang, Y., Luo, E., 2024. Understanding the Barriers to Consumer Purchasing of Electric Vehicles: The Innovation Resistance Theory. Sustainability 16 (6), 2420.

Yang, H., Lee, H., 2019. Understanding user behavior of virtual personal assistant devices. Information Systems and e-Business Management 17, 65–87.

Yang, H., Lee, H., Zo, H., 2017a. User acceptance of smart home services: an extension of the theory of planned behavior. Industrial Management & Data Systems 117 (1), 68–89.

Yang, Q., Pang, C., Liu, L., Yen, D.C., Michael Tarn, J., 2015. Exploring consumer perceived risk and trust for online payments: An empirical study in China's younger generation. Computers in Human Behavior 50, 9–24.

Yang, S., Song, Y., Tong, S., 2017b. Sustainable Retailing in the Fashion Industry: A Systematic Literature Review. Sustainability 9 (7), 1266.

Yang, S., Xiong, G., 2019. Try It On! Contingency Effects of Virtual Fitting Rooms. Journal of Management Information Systems 36 (3), 789–822.

Yang, X., Chen, S.-C., Zhang, L., 2020. Promoting sustainable development: A research on residents' green purchasing behavior from a perspective of the goal-framing theory. Sustainable Development 28 (5), 1208–1219.

Yew, G.C.K., 2021. Trust in and Ethical Design of Carebots: The Case for Ethics of Care. Int J of Soc Robotics 13 (4), 629–645.

Zalando, 2023. Zalando brings a virtual fitting room pilot to millions of customers <https://corporate.zalando.com/en/technology/zalando-brings-virtual-fitting-room-pilot-millions-customers> (accessed 1 July 2024).

Zavolokina, L., Zani, N., Schwabe, G., 2019. Why Should I Trust a Blockchain Platform? Designing for Trust in the Digital Car Dossier. International Conference on Design Science Research in Information Systems and Technology, 269–283.

Zhang, B., Anderljung, M., Kahn, L., Dreksler, N., Horowitz, M.C., Dafoe, A., 2021. Ethics and Governance of Artificial Intelligence: Evidence from a Survey of Machine Learning Researchers. Journal of Artificial Intelligence Research 71, 591–666.

Zhang, D., 2023. Environmental Sustainability and Global Value Chains: Challenges and Opportunities for Green Supply Chain Management. Frontiers in Business, Economics and Management 10 (2), 155–160.

Zhao, B., Rawwas, M.Y., Zeng, C., 2020. How does past behaviour stimulate consumers' intentions to repeat unethical behaviour? The roles of perceived risk and ethical beliefs. Business Ethics: A European Review 29 (3), 602–616.

Zhao, X., Lynch, J.G., Chen, Q., 2010. Reconsidering Baron and Kenny: Myths and Truths about Mediation Analysis. Journal of Consumer Research 37 (2), 197–206.

Zhu, P., Wang, Z., Li, X., Liu, Y.-H., Zhu, X., 2020. Understanding promotion framing effect on purchase intention of elderly mobile app consumers. Electronic Commerce Research and Applications 44, 101010.

Zinkann, R., Mahadevan, J., 2018. Zukünftige Customer Journeys und deren Implikationen für die Unternehmenspraxis, in: , Marketing Weiterdenken. Springer Gabler, Wiesbaden, 157–169.

Zucker, L.G., 1986. Production of trust: Institutional sources of economic structure. Research in Organizational Behavior 8 (1), 53–111.

The manufacturer's authorised representative in the EU is Springer Nature Customer Service Centre GmbH, Europaplatz 3, 69115 Heidelberg, Germany. If you have any concerns regarding our products, please contact ProductSafety@springernature.com

Printed and bound by CPI Group (UK) Ltd, Croydon, CR0 4YY

02/10/2026

02227140-0002